Voice Technologies for Speech Reconstruction and Enhancement

Speech Technology and Text Mining in Medicine and Health Care

Edited by
Amy Neustein

Volume 6

Published in the Series

Patil, Neustein, Kulshreshtha (Eds.),
Signal and Acoustic Modeling for Speech and Communication Disorders, 2018
ISBN 978-1-61451-759-7, e-ISBN 978-1-5015-0241-5,
e-ISBN (EPUB) 978-1-5015-0243-9

Ganchev, *Computational Bioacoustics*, 2017
ISBN 978-1-61451-729-0, e-ISBN 978-1-61451-631-6, e-ISBN (EPUB) 978-1-61451-966-9

Beals et al., *Speech and Language Technology for Language Disorders*, 2016
ISBN 978-1-61451-758-0, e-ISBN 978-1-61451-645-3, e-ISBN (EPUB) 978-1-61451-925-6

Neustein (Ed.), *Speech and Automata in Healthcare*, 2014
ISBN 978-1-61451-709-2, e-ISBN 978-1-61451-515-9,
e-ISBN (EPUB) 978-1-61451-9607, Set-ISBN 978-1-61451-516-6

Neustein (Ed.), *Text Mining of Web-Based Medical Content*, 2014
ISBN 978-1-61451-541-8, e-ISBN 978-1-61451-390-2,
e-ISBN (EPUB) 978-1-61451-976-8, Set-ISBN 978-1-61451-391-9

Voice Technologies for Speech Reconstruction and Enhancement

Edited by
Hemant A. Patil, Amy Neustein

DE GRUYTER

Editors
Dr. Hemant A. Patil
ISCA Distinguished Lecturer 2020-2021
APSIPA Distinguished Lecturer 2018-2019
Room No. 4103, Faculty Block-4
Near Indroda Circle
DA-IICT Gandhinagar
382 007, Gujarat State, India
hemant_patil@daiict.ac.in

Dr. Amy Neustein
Linguistic Technology Systems
1530 Palisade Avenue, Suite 28R
Fort Lee NJ 07024
USA
amy.neustein@verizon.net

ISBN 978-1-5015-2666-4
e-ISBN (PDF) 978-1-5015-0126-5
e-ISBN (EPUB) 978-1-5015-0130-2
ISSN 2329-5198

Library of Congress Control Number: 2019948019

Bibliographic information published by the Deutsche Nationalbibliothek
The Deutsche Nationalbibliothek lists this publication in the Deutsche Nationalbibliografie;
detailed bibliographic data are available on the Internet at http://dnb.dnb.de.

© 2021 Walter de Gruyter Inc., Boston/Berlin
This volume is text- and page-identical with the hardback published in 2020.
Cover image: MEHAU KULYK/SCIENCE//PHOTO LIBRARY/Agentur Focus
Typesetting: Integra Software Services Pvt. Ltd.
Printing and binding: CPI books GmbH, Leck

www.degruyter.com

Foreword

This book is a collection of seven chapters from different renowned research groups from across the world, all dealing with improving speech from speakers with medical disorders, primarily dysarthria.

The work relates to the analysis, processing and improvement of such speech. Dysarthria is a neuromuscular speech disorder, often resulting from neurological injury. It can be developmental or acquired. Dysarthria may affect respiration, phonation, resonance, articulation and prosody. The physical production of speech is impaired: poor articulation, uneven volume, variable speech rate, incorrect pitch, errors in voicing and/or slurring. Dysarthric speech is less comprehensible than normal speech.

The first of three parts examines three application areas for disordered speech, namely, speaker recognition, enhancement and intelligibility.

Identifying people by their voice can be effective and efficient, given the cost of specialized equipment for other biometrics (e.g., retinal scans, fingerprints). However, the accuracy of speaker recognition depends on the condition of the speech, for example, pathological speech. Chapter 1 examines such effects for dysarthric speech, exploring standard representations, that is, i-vectors, bottleneck neural networks and covariance-based features. The TORGO and NEMOURS databases serve for evaluation.

Chapter 2 deals with ways to help people with this disorder by enhancement of disordered speech, that is, methods to transform dysarthric speech into more natural and intelligible speech, while retaining style characteristics of the speaker. Unlike some systems that require the speaker to place a pause after each word, the work here allows continuous speech. Given the multiple causes and types of dysarthria, it was a challenge to develop a generic technique to enhance many types of dysarthric speech. Emphasis is placed on durational modifications. Performance is compared with two other techniques: resynthesis via formants or via hidden Markov model. An Indian English dysarthric speech dataset was used in the experiments.

The third chapter examines assessment and intelligibility modification for dysarthric speech.

It is noted that dysarthric speech may have unusual breaks in pitch, voice tremor, monotonic loudness or excessive loudness variation and flat pitch. There can also be hypernasality due to velo-pharyngeal incompetence, irregular articulatory breakdown, repetition of phonemes and distortion of vowels due to a defective articulatory system. There are wide variations in pitch jitter and amplitude shimmer, as well as intensity variations. These can cause additional resonances or antiresonances in the speech, or widening of formant

https://doi.org/10.1515/9781501501265-202

bandwidths, which reduce the quality of the speech. This chapter looks at acoustic analysis for assessment of dysarthric speech as well as successful approaches for intelligibility enhancement.

The second of the three book parts describes reconstruction of dysarthric speech from sources other than speech, that is, from a physiological microphone (PMIC) and from murmur. These can be useful for communication of a received audio signal where speech is absent or heavily distorted.

Chapter 4 examines the PMIC, a nonacoustic contact sensor. Due to the alternative pickup location of the PMIC, the audio signal sounds muffled and metallic, but has variations appropriate to the speaker. This chapter presents a probabilistic transformation approach to improve the perceptual quality and intelligibility of PMIC speech. Performance is evaluated objectively and subjectively.

Chapter 5 examines how a speaker's murmur can be transformed into useful speech. This is a silent speech interface technique, using a nonaudible microphone (NAM). For speakers with a vocal fold disorder, movements of speech articulators can be estimated through the soft tissue of the head using a NAM microphone, which gets a weak nonaudible murmur. The soft tissue suppresses most of the higher-frequency information that speech normally has. This work uses a deep neural network-based conversion technique with the rectifier linear unit as its nonlinear activation function. It has a two-stage model where first stage converts the NAM signal to whispered speech and a second model converts to audible speech. Performance is compared with a state-of-the-art GMM-based method (NAM2WHSP). The second stage part also helps to extract linguistic message content from the NAM signal.

The last of the book parts describes intervention software for these applications of dysarthric speech.

Chapter 6 deals with ways to have early identification of communication disorders, as well software-based therapy solutions to improve communication skills. This chapter gives an overview of communication disabilities, the parameters for a speech production model, a brief introduction to speech analysis and synthesis with communication disabilities and an overview of Vagmi and related software.

Chapter 7 describes a way to perform speech therapy remotely, that is, via telephony – the mobile phone-assisted remote speech therapy (MoPAReST) platform. Traditional rehabilitation for speech disabilities requires physical or virtual real-time interactions. Recent work between speech language pathologists (SLP) and speech technologists has led to improved tools for speech rehabilitation. The MoPAReST platform consists of a web interface for the SLP to monitor the progress of their patients and a mobile application for the patient to practice therapy exercises. This allowed digitization of the usual paper-based assessments and therapy drills, personalization for different target audiences, algorithms for objective assessment and evaluation of the platform with five actual patients.

This book is particularly relevant today, as healthcare systems increasingly use ubiquitous and pervasive computing to monitor patients at a distance. The editors of this volume have done a fine job in reviewing recent international work in this field. They have presented here current and advanced research findings in signal and acoustic modeling of speech pathologies. The target audience for this book is speech scientists, clinicians and pathologists.

Prof. Douglas O'Shaughnessy
(Ph.D. MIT USA, IEEE Fellow and ASA Fellow)
INRS-EMT, University of Quebec, and McGill University, Canada

Acknowledgments

This coedited book volume is furthered by the stellar contributions of authors across the globe, including two IEEE fellows, Prof. John H. L. Hansen and Prof. Douglas O'Shaughnessy, the latter of whom wrote the foreword. Special thanks to Prof. T. V. Ananthapadmanabha for writing a comprehensive and reflective solo chapter drawn from his wealth of clinical experience and his pioneering approach to assessing and treating speech and communication disorders.

In addition, special thanks to the authorities of DA-IICT Gandhinagar and the members of Speech Research Lab at DA-IICT, whose steadfast encouragement has sustained us throughout this lengthy project. Special thanks are due to series editor, Dr. Amy Neustein, whose patience, trust and confidence in the research capabilities of both the authors and the coeditor helped bring this challenging book project to fruition. Finally, we acknowledge the generous support, cooperation and patience from the editorial team and production staff at De Gruyter, and in particular Leonardo Milla, who guided us properly and made this book possible.

Prof. Hemant A. Patil, DA-IICT Gandhinagar, Gujarat State, India.
ISCA Distinguished Lecturer 2020–2021
APSIPA Distinguished Lecturer (DL) 2018–2019

https://doi.org/10.1515/9781501501265-203

Contents

Part I: **Comparative analysis of methods for speaker identification, speech recognition, and intelligibility modification in the dysarthric speaker population**

Part II: **New approaches to speech reconstruction and enhancement via conversion of non-acoustic signals**

Part III: **Use of novel speech diagnostic and therapeutic intervention software for speech enhancement and rehabilitation**

List of contributors

T.V. Ananthapadmanabha
Voice and Speech Systems,
India
tva.blr@gmail.com

Chitralekha Bhat
TCS Research and Innovation,
India
bhat.chitralekha@tcs.com

Patrick Cardinal
École de technologie supérieure ÉTS,
Canada
patrick.cardinal@etsmtl.ca

M. Dhanalakshmi
SSN College of Engineering,
India
dhanalakshmim@ssn.edu.in

Tiago H. Falk
Institut National de la Recherche
Scientifique, Centre Énergie, Matériaux,
Télécommunications and MuSAE Lab,
Canada
falk@emt.inrs.ca

John H. L. Hansen
Center for Robust Speech Systems (CRSS),
University of Texas at Dallas,
USA
john.hansen@utdallas.edu

Anjali Kant
Ali Yavar Jung National Institute for Speech
and Hearing Disabilities,
India
ayjnihh-mum@nic.in

Sunil Kumar Kopparapu
TCS Research and Innovation,
India
sunilkumar.kopparapu@tcs.com

François Michaud
Université Sherbrooke,
Canada
francois.michaud@usherbrooke.ca

Hema A. Murthy
Indian Institute of Technology Madras,
India
hema@iitm.ac.in

T. Nagarajan
SSN College of Engineering,
India
nagarajant@ssn.edu.in

Seyed Omid Sadjadi
Center for Robust Speech Systems (CRSS),
University of Texas at Dallas,
USA
sadjadi@utdallas.edu

Hemant A. Patil
Dhirubhai Ambani Institute of Information
and Communication Technology (DA-IICT),
India
hemant_patil@daiict.ac.in

Sanjay A. Patil
Center for Robust Speech Systems (CRSS),
University of Texas at Dallas,
USA
sanjay.patil@utdallas.edu

Anusha Prakash
Indian Institute of Technology Madras,
India
anushaprakash90@gmail.com

M. Ramasubba Reddy
Indian Institute of Technology Madras,
India
rsreddy@iitm.ac.in

https://doi.org/10.1515/9781501501265-205

Sarita Rautara
Ali Yavar Jung National Institute for Speech
and Hearing Disabilities,
India
ayjnihh-mum@nic.in

Milton Sarria-Paja
Universidad Santiago de Cali,
Colombia
milton.sarria00@usc.edu.co

Mohammed Senoussaoui
École de technologie supérieure (ÉTS)
and Fluent.ai,
Canada
mohammed.senoussaoui.1@etsmtl.net

Nirmesh J. Shah
Dhirubhai Ambani Institute of Information
and Communication Technology (DA-IICT),
India
nirmesh88_shah@daiict.ac.in

Bhavik Vachhani
ScribeTech and TCS Research and
Innovation,
India
bhavik.v@scribetech.in

P. Vijayalakshmi
SSN College of Engineering,
India
vijayalakshmip@ssn.edu.in

Introduction

Voice Technologies for Speech Reconstruction and Enhancement presents under one heading some of the most advanced methods and techniques for reconstructing and enhancing dysarthric speech. Using empirically sound and reliable testing procedures, the authors show how their new methods compare with standard approaches for improving speech that is compromised by various neuromotor disorders, collectively known as "dysarthria" – a word that originates in "dys" and "arthrosis," which means difficult or imperfect articulation. Such neuromuscular speech disorders are the consequences of neurological injury: they can be developmental as in cerebral palsy or acquired as in Parkinson's disease, traumatic brain injury or a stroke.

In examining new voice technologies, contributors to this volume explore, among other things, novel ways to improve speaker biometrics among dysarthric speakers. They test advances in speech recognition research under dysarthric speech, where such advances had not been used heretofore for identifying speakers suffering from dysarthria. Among their suite of tools used to perform speaker recognition in the dysarthric population, contributors creatively utilize covariance-based features, which have been traditionally used for emotion classification but not for speaker identification. Here for the first time, covariance-based features are explored within the realm of speaker recognition, particularly for dysarthric speakers.

As a point in fact, while speech recognition in the dysarthric population has inspired much research, automatic speaker (and not speech) recognition among the population of dysarthric speakers has not received sufficient attention. The authors in this volume address these lacunae, showing how new methods can improve speaker biometrics when speech is impaired due to developmental or acquired pathologies.

Contributors emphasize the importance of creating aids that ensure that "more natural and intelligible speech is produced" while at the same time "retaining the characteristics of the speaker." To achieve this goal, contributors propose as a first step: the exploration of the "durational attributes across dysarthric and normal speech utterances," using an automatic technique that was developed by the authors for the specific purpose of correcting dysarthric speech, so as to bring it closer to normal speech.

The authors take a close look at the design of speech systems and offer some novel approaches to perfecting the design of the system. Starting from the premise that dysarthria affects speech systems globally, the authors remain very much attuned to the fact that speech systems usually have trouble in

https://doi.org/10.1515/9781501501265-001

differentiating the *multiple* dimensions of dysarthric speech. To meet this challenge head on, the authors offer a novel multidimensional approach to help the speech system both assess dysarthric speech and to perform intelligibility improvement of the impaired speech. In so doing, they factor in laryngeal, velopharyngeal and articulatory subsystems for dysarthric speakers, by using a speech recognition system and relevant signal-processing-based techniques. The authors show how by using results derived from the assessment system, dysarthric speech can be corrected and resynthesized. This successfully preserves the speaker's identity and greatly improves overall intelligibility.

The contributors to this anthology, in providing an incisive look at new approaches to speech reconstruction and enhancement, consider nonacoustic signals and muted nonverbal sounds in relation to audible speech conversion. They describe, for example, the nonaudible microphone (NAM) for patients who suffer from vocal fold disorders, where vocal fold vibrations are absent. In using a novel approach of deep neural network-based conversion technique with a rectifier linear unit as a nonlinear activation function, they describe how one is able to "convert less intelligible NAM signal to an audible speech signal." This is particularly useful where high-frequency-related information is attenuated by lack of radiation effect at the lips and other structural issues regarding soft tissue.

Equally important, the contributors explore ways to improve the physiologic microphone (PMIC), a nonacoustic contact sensor that provides an alternative capture signal where there is severe vocal fold pathology. They present empirical results of "a probabilistic transformation approach to improve the perceptual quality and intelligibility of PMIC speech by mapping the nonacoustic signal into the conventional close-talk acoustic microphone speech production space, as well as by minimizing distortions arising from alternative pickup location." They show how this technique facilitates "more robust and natural human-to-human speech communication."

Taking an impassioned look into the best technologies for assisting dysarthric speakers, including those that reside in under-resourced communities with severely limited access to speech therapists, contributors to this volume present their own innovative software solutions that have withstood the rigor of clinical trials. Using novel speech diagnostic and therapeutic intervention software in underdeveloped regions, the authors show how speech can be enhanced notwithstanding the scarcity of speech therapists. Paying close attention to proper testing and evaluation, contributors amply demonstrate how software solutions perform in redressing developmental and acquired speech impairments and in redressing vocal injuries stemming from overuse of a normal voice, as is common in teaching, singing and campaigning for public office.

This anthology brings together researchers, academicians and hands-on clinicians from the United States and Canada and from the high-tech epicenters in various regions in India. This book is divided into three parts.

The book opens with "Comparative Analysis of Methods for Speaker Identification, Speech Recognition and Intelligibility Modification in the Dysarthric Speaker Population." This part examines specific methods for improving both speaker and speech recognition capabilities of current speech systems when dealing with dysarthric speech. Topics include: 1) evaluating the best performing speaker recognition systems among dysarthric speakers; 2) developing assistive speech technologies for the population of dysarthric speakers so as to improve the naturalness and intelligibility of dysarthric speech while retaining the specific characteristics of the speaker and 3) demonstrating the empirical performance of a special assessment system where dysarthric speech is corrected and resynthesized, preserving speaker identity and improving intelligibility.

The second part, "New Approaches to Speech Reconstruction and Enhancement via Conversion of Nonacoustic Signals and Nonaudible Murmur," provides an incisive look at some leading edge techniques for maximizing the use of nonacoustic signals and muted nonverbal sounds in patients who suffer from such severe vocal fold disorders that the routine vocal fold vibrations are absent in those patients. They study the performance of the NAM and the PMIC, showing how such techniques are effective in converting nonacoustic speech signals and nonaudible murmurs to intelligible speech so as to improve human-to-human communication and reduce the difficulty of the dysarthric speaker who struggles continually to be heard and understood by others.

The third part, "Use of Novel Diagnostic and Therapeutic Intervention Software for Speech Enhancement and Rehabilitation," rounds out this volume by presenting the empirical results of novel speech diagnostic and therapeutic intervention software. This software, which can be used in developed regions, is of particular importance in under resourced communities where there is limited access to speech therapists. One of the contributors to this part provides a fascinating presentation of the "Vagmi" software solution for assessing and treating voice and speech disorders, showing how this novel software solution works for both voice diagnostics and for therapeutic intervention. Following that, there is a discussion of the "Mobile Phone-Assisted Remote Speech Therapy (MoPAReST) Platform." This platform provides patients with speech therapy remotely when there are minimal face-to-face sessions with speech therapists. Filling in major lacunae in the delivery of speech therapy services, this platform provides an interesting web interface for the speech pathologist to monitor the patient at any time round the clock, and a user-friendly mobile application for the patient to practice the therapy exercises at a time and location of their convenience.

As editors of this volume, we have strived to fill an important gap is signal-processing and speech research. As part of the de Gruyter Series in *Speech Technology and Text Mining in Medicine and Health Care*, this book provides a candid and open discussion of methods and techniques for reconstructing and enhancing dysarthric speech. As part of our discussion we emphasize empirical research demonstrating how speech systems can boost their automatic speech/speaker recognition accuracy rates even when speech input is severely compromised by articulatory problems. We encourage readers of *Voice Technologies for Speech Reconstruction and Enhancement* to partake in the discussion and build an expanding corpus of knowledge and research, using the empirical findings in this volume as a springboard for further research and testing. Undoubtedly, dysarthric speakers deserve better performing speech and speaker recognition systems that accommodate to the manifold speech impairments under the broad heading of dysarthria. We are indeed grateful to the contributors for sharing their research on new voice technologies for speech reconstruction and enhancement so that those with speech disorders are not left on the sidelines, struggling to be understood both by humans and by machines. As scientists, we join hands with those who are most vulnerable to social prejudices due to speech and language disorders, offering assistive voice technologies to reconstruct and enhance their speech so that the avenues of communication will once again be open to them.

Part I: **Comparative analysis of methods for speaker identification, speech recognition, and intelligibility modification in the dysarthric speaker population**

Mohammed Senoussaoui, Milton O. Saria-Paja,
Patrick Cardinal, Tiago H. Falk and François Michaud

1 State-of-the-art speaker recognition methods applied to speakers with dysarthria

Abstract: Speech-based biometrics is one of the most effective ways for identity management and one of the preferred methods by users and companies given its flexibility, speed and reduced cost. Current state-of-the-art speaker recognition systems are known to be strongly dependent on the condition of the speech material provided as input and can be affected by unexpected variability presented during testing, such as environmental noise, changes in vocal effort or pathological speech due to speech and/or voice disorders. In this chapter, we are particularly interested in understanding the effects of dysarthric speech on automatic speaker identification performance. We explore several state-of-the-art feature representations, including i-vectors, bottleneck neural-network-based features, as well as a covariance-based feature representation. High-level features, such as i-vectors and covariance-based features, are built on top of four different low-level presentations of dysarthric/controlled speech signal. When evaluated on TORGO and NEMOURS databases, our best single system accuracy was 98.7%, thus outperforming results previously reported for these databases.

Keywords: speaker recognition, dysarthria, i-vectors, covariance features, bottleneck features

1.1 Introduction

Human speech is a natural, unique, complex and flexible mode of communication that conveys phonological, morphological, syntactic and semantic information

Mohammed Senoussaoui, Patrick Cardinal, École de technologie supérieure (ÉTS) and Fluent.ai, Canada
Milton O. Saria-Paja, Universidad Santiago de Cali, Colombia
Tiago H. Falk, Institut National de la Recherche Scientifique, Centre Énergie, Matériaux, Télécommunications and MuSAE Lab, Canada
François Michaud, Université Sherbrooke, Canada

https://doi.org/10.1515/9781501501265-002

provided within the utterance [1]. It also conveys traits related to identity, age, emotional or health states, to name a few [2, 3]. This information has been useful across a number of domains; for example, automatic speech recognition (ASR) has opened doors for speech to be used as a reliable human–machine interface [4, 5]. Advances in speaker recognition (SR) technologies, in turn, have allowed humans to use their voice to, for example, authenticate themselves into their bank's automated phone system [6]. Under most circumstances, speech-enabled applications have been developed to work in clean environments and assume clear and normal adult speech. These assumptions and conditions, however, are difficult to satisfy in many real-world environments. Speech production requires integrity and integration of numerous neurological and musculoskeletal activities. Many factors, such as accidents or diseases, however, can affect the quality and intelligibility of produced speech [7, 8]. These modifications are usually referred to as speech disorders, and their effects can be observed in individuals of varying age groups for different causes. Dysarthria is a particular speech motor disorder caused by damage to the nervous system and characterized by a substantive decrease in speech intelligibility [9–11]. Reduced intelligibility can negatively influence an individual's life in several ways, including social interactions, access to employment, education or interaction with automated systems [12]. Many speech-enabled applications have thrived due to the recent proliferation of mobile devices. Notwithstanding, while the ubiquity of smartphones has opened a pathway for new speech applications, it is imperative that the performance of such systems be tested for pathological speech, such that corrective measures can be taken, if needed.

At present, most of the research conducted in the digital speech-processing domain applied to speech disorders has focused mainly on ASR, enhancement and speech intelligibility assessment [13–15]. Other emerging speech applications have yet to be explored, such as SR, language identification, emotion recognition, among others. Particularly, in this chapter the SR problem is of special interest. Such technologies are burgeoning for identity management as they eliminate the need for personal identification numbers, passwords and security questions [16]. In this regard, Gaussian mixture models (GMMs) combined with mel-frequency cepstral coefficients (MFCC) as feature vectors, known as the GMM-MFCC paradigm, were for many years the dominant approach for text-independent SR [6]. Over the last decade, the i-vector feature representation [17] has become the state-of-the-art for text-independent SR, as well as for many other speech-related fields such as language recognition [18–21] and ASR [22]. Recently, the i-vector framework was also successfully applied to objective dysarthria intelligibility assessment [23].

Covariance-based features, on the other hand, were first proposed for the task of object detection [24] and were further explored for object detection and tracking [25, 26]. In speech-processing applications, covariance-based features have been successfully used for emotion classification from speech [27]. Here, covariance-based features will be explored for the first time within the scope of SR, in particular for dysarthric speakers. Both the i-vector representation and the covariance-based features are techniques to map variable length frame-based feature representations to a fixed-length feature vector while retaining most relevant information from the low-level or frame-based feature representation. Henceforth, we will refer to this class of features as utterance-based.

To the best of our knowledge, there is only one published work exploring SR for dysarthric speakers [28]. In [28], promising results for speaker identification (SID) were found by fusing MFCCs with auditory cues-based features and by using traditional classification methods, such as GMM and support vector machines (SVM), directly on the above low-level features. However, the above-mentioned advances in SR research have not been tested yet under dysarthric speech. Hence, it is not clear what are the performance boundaries achievable with dysarthric speech on existing state-of-the-art systems. Here, we aim to fill this gap.

More specifically, in this work, the i-vector and covariance-based feature representation will be built on top of four different frame-based features, which are expected to capture short- and long-term temporal speech dynamics. For this purpose, the classical MFCC features were chosen, as they provide an efficient speech spectrum representation that incorporates some aspects of human audition and have been useful in numerous applications using pathological speech [2, 29–30]. Moreover, the slowly varying temporal envelope of bandpass speech signals has also shown to contain highly discriminative speaker-dependent information [6, 31, 32]. Here, two long-term temporal dynamics signal representations are explored: the first relies on the AM-FM representation proposed in [32] and the second on the modulation spectrum (MF) derived from the short-time Fourier transform, as described in [33]. Lastly, current state-of-the-art SR systems rely on the extraction of i-vectors from the so-called bottleneck features (BNF) [34], which have replaced the classical MFCC as acoustic features using an approach based on deep learning. Robustness of these approaches has yet to be tested for pathological speech.

The remainder of this chapter is organized as follows. Section 2 details the frame- and utterance-level speech representations used herein. Speech material, experimental protocol and technical details are then presented in Section 3. Experimental results and discussions are given in Section 4 and conclusions and future research directions in Section 5.

1.2 Speech signal representations

Speech is produced from a time-varying vocal tract system, which makes speech signals dynamic or time varying in nature. Even though the speaker has control over many aspects of speech production, for example, loudness, voicing, fundamental frequency and vocal tract configuration, much speech variation is not under speaker control, for example, periodicity of the vocal fold vibration [2]. Dysarthric speakers, on the other hand, have much less control over the vocal tract system, thus causing major changes in the acoustic and the general dynamic characteristics of the speech signal when compared with control or healthy speakers [14]. Since speech is inherently a nonstationary signal, in order to be able to use many of the available mathematical tools, short-time duration blocks need to be used, where the signal satisfy stationary conditions. Such "short-time" processing methods, as they are known, can be performed either in the time or in the frequency domain [2, 35] and have been for many years the standard approach to characterize speech signals.

In this work, four different frame-based, or short-time, features are evaluated, namely, MFCC, AM-FM features, MF and DNN BNFs, aiming at capturing both short- and long-term temporal speech dynamics. Typically, for speech-processing applications, additional postprocessing is performed on the resulting short-time features to mitigate the effects of linear channel mismatch, and to add robustness to the overall system. Standard techniques include short-time mean and variance normalization, short-time mean and scale normalization and short-time Gaussianization [36], which are typically applied using a sliding window of 3 s over the short-time variables. However, given the short duration of the recordings used in our experiments (average length between 3 and 5 s), performing such a short-time processing becomes nonviable. Therefore, we adopt a per-utterance scheme to carry out cepstral mean normalization (CMN) or cepstral mean and variance normalization (CMVN). Moreover, on top of these frame-based features, two high-level (i.e., utterance-based) representations, namely, i-vector and covariance features (CFs) are also explored.

1.2.1 Frame-based features

1.2.1.1 Mel-frequency cepstral coefficient (MFCC)

For MFCC computation, and to emulate human cochlear processing, 20 triangular bandpass filters spaced according to the mel scale are imposed on the power spectrum. Next, a discrete cosine transform is applied to the log-filterbank outputs, to obtain a 20-dimensional feature vector, including the 0th order cepstral coefficient (log energy). The temporal changes in adjacent frames play a significant role in human perception; thus to capture this dynamic information in the speech, dynamic or transitional features (Δ and $\Delta\Delta$ MFCC) are appended to the feature vector, such that the final feature vector is 60-dimensional. These features are computed on a per-window basis using a 20 ms window with 50% overlap, and a sampling rate of 16 kHz is used.

1.2.1.2 AM-FM-based features

For speech analysis using the AM-FM signal representation, it is assumed that an observed time-domain signal is the result of multiplying a low-frequency modulator (temporal envelope) by a high-frequency carrier. Hence, the MF characterizes the rate of change of long-term speech temporal envelopes [30], and the analysis is carried out by using acoustic subbands. Thus, in this approach the AM-FM model decomposes the speech signal into bandpass channels and characterizes each channel in terms of its envelope and phase (instantaneous frequency) [32]. Features derived from the AM-FM signal representation have proven to be more robust in noisy conditions and perform at the same level as cepstral coefficients in SID tasks [31, 32].

For feature extraction purposes, the speech signal $x(n)$ is filtered through a bank of H_K filters, resulting in K bandpass signals: $y_k(n) = x(n) * h_k(n)$, where $h_k(n)$ corresponds to the impulse response of the kth filter. There are different approaches for filter design that have been used in speech applications [32]. In our experiments, a 27-channel gammatone filterbank is used, with filter center frequencies (fc_k) ranging from 100 Hz to 7000 Hz and bandwidths characterized by the mel scale. It was concluded from a pilot experiment that this is an optimal setting for our purposes. After filtering, each analytic subband signal $x_k(n)$ is uniquely related to a real-valued bandpass signal $y_k(n)$ by $x_k(n) = y_k(n) + j \cdot \hat{y}_k(n)$, where $\hat{y}_k(n)$ stands for Hilbert transform of $x_k(n)$ and j is the imaginary unit.

For the sake of notation, let $a_k(n)$ denote the low-frequency modulator and $f_k(n)$ the instantaneous frequency for each bandpass signal. Then, the weighted instantaneous frequencies (WIF) are computed by combining the values of $a_k(n)$ and $f_k(n)$ using a short-time approach, using a 25 ms window with 40% overlap, more specifically:

$$F_k = \frac{\sum_{i=n_0}^{n_0+\tau} f_k(i) \cdot a_k^2(i)}{\sum_{i=n_0}^{n_0+\tau} a_k^2(i)}, \quad k = 1, \ldots, 27, \tag{1.1}$$

where n_0 is the starting sample point and τ is the length of the time frame. Pre-emphasis and feature normalization were not required and WIF features are expressed in kHz, as suggested by [32].

1.2.1.3 Modulation spectrum (MS)

Features such as WIF use a time domain approach to extract information from subband speech signals. The modulation frequency (modulation domain) represents the frequency content of the subband amplitude envelopes and it potentially contains information about speaking rate and other speaker-specific attributes [6]. Auditory-inspired amplitude modulation features have been effectively used in the past to improve automatic SID in the presence of room reverberation [37] and speaking style classification in noisy environments [38]. In both cases, the environment-robustness property of the modulation spectral features was observed. To compute the auditory-inspired amplitude modulation features, the approach presented in [33] was used. Herein, we present a complete description of the algorithm for the sake of completeness. The speech signal $x(n)$ is first processed by an N-point short-time discrete Fourier transform to generate $X(nL, f_a)$ given by

$$X(nL_a, f_a) = \sum_{m=-\infty}^{\infty} X(m)\omega_a(nL_a - m)e^{-i\frac{2\pi k}{N}m}, \tag{1.2}$$

where $\omega_a(n)$ is an acoustic frequency analysis window, the subscript a stands for acoustic domain and L_a denotes the frame shifts. In order to emulate human cochlear processing, the squared magnitudes of the obtained acoustic frequency components are grouped into 27 subbands ($|X_k(\cdot)|, k = 1, \ldots, 27$), spaced according to the perceptual mel scale. A second transform is then performed across time for each of the 27 subband magnitude signals to yield:

$$X_j(mL_m, f_m) = \sum_{n=-\infty}^{\infty} |X_j(n)|\omega_m(mL_m - n)e^{-j\frac{2\pi k}{N}n}, \tag{1.3}$$

where $\omega_m(m)$ is a modulation frequency analysis window, the subscript m stands for modulation domain, L_m the frame shift and j indexes the acoustic frequency bands. We further group squared modulation frequency bins into eight subbands using logarithmically spaced triangular bandpass filters distributed between $0.01 - 80$ Hz modulation frequency, as motivated by recent physiological evidence of an auditory filterbank structure in the modulation domain [39]. The speech MF results in a high-dimensional feature representation (e.g., 27 acoustic bands $\times 8$ modulation bands $= 216$ dimensions); finally log_{10} compression is applied. Each recording is represented as a three-dimensional array, with dimensions being acoustic frequency, modulation frequency and time. Each time context (a matrix with 216 elements) can be collapsed into a vector and used as standard features. However, given the relative high dimensionality of the resulting space and correlation among different dimensions, each feature vector is projected to a lower-dimensional space using principal component analysis (PCA) with 40 components retaining 98.7% of cumulative variance, which according to our experiments showed to be an optimum value. The differences of this approach with the one presented in [40] are the way the modulation bins are group in logarithmic distributed bands, the log compression and finally the dimensionality reduction, which according to our experiments result in a more informative feature vector.

1.2.1.4 Bottleneck features (BNF)

Bottleneck neural network are deep neural networks (DNN) with a particular topology where one of the hidden layers has significantly lower dimensionality than the surrounding layers; such a layer is known as the *bottleneck layer*. A BNF vector is obtained by forwarding a primary input feature vector through the DNN and reading off the vector of values at the bottleneck layer [5, 34]. For SR purposes, the DNN for feature extraction needs to be trained first for a specific frame-by-frame classification task. According to previous reports, excellent results were observed with features extracted using DNN trained for a phone-like classification, specifically if the targets are subphonetic units known as "senones" [34].

In practice, it is difficult to establish a direct connection between how the human auditory system works and the role of weights and non linear operations in a DNN to treat input information to simulate the auditory system. Consequently, for the task at hand, it is difficult to set any specific architecture aiming to better describe the type of data we want to characterize. In this regard, recently it was reported that the position of the bottleneck layer has to do with the task at hand and how similar the data used for training the DNN is to the evaluation set. McLaren et al. [41] suggested to use the bottleneck layer close to the DNN output layer when the training data is matched to the evaluation conditions, and a layer more central to the DNN otherwise. Therefore, the adopted DNN architecture is depicted in Figure 1.1. The configuration and the training conditions allow us to evaluate the robustness of these features in atypical scenarios, and evaluate how efficiently a system trained with these features can generalize to other speech modalities such as pathological speech.

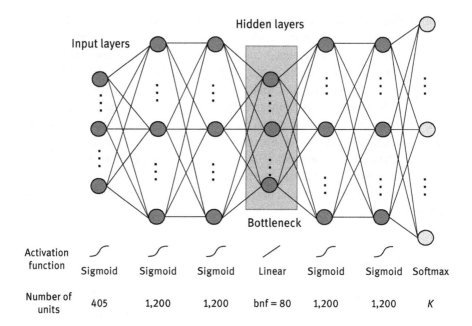

Figure 1.1: Bottleneck neural-network architecture used in this work. *K* is the number of targets.

Furthermore, for the task at hand, data insufficiency is one of the major problems in building an efficient speech-enabled system capable of handling both normal and pathological speech. The solution taken in this work is to use relatively large amounts of normally phonated speech for parameter estimation, as done before for ASR [42], and explore highly informative features containing speaker-dependent information, and, if possible, independent of the speech modality. Atypical speech, including dysarthric speech, has been shown to contain high variability at lower acoustic bands, mainly in frequency ranges containing information related to first and second formants [43, 44]. In this regard, in previous work exploring whispered speech speaker verification, it was suggested that by using the subband from 1.2 kHz to 4 kHz to compute the different feature sets it was possible to improve performance in the mismatch condition, but at the cost of reduced performance in the matched scenario [45, 33]. Taking this into account, two different approaches to compute BNFs are explored in this work:

    BNF 1: In this approach, the targets for the DNN were obtained using a CD-GMM-HMM (context dependent–hidden Markov model using GMMs to model observations) ASR system trained with Kaldi [46]. In total 460 h of recordings were used as DNN training data; the recordings were extracted from the LibriSpeech dataset [47]. The speech recordings were character-ized on a per-window basis using 27 log Mel-scale filter bank outputs, with 25 ms frames and 10 ms hop time. The DNN input features are concatenated time contexts of 15 frames, which results in a 405-dimensional vector. As previously mentioned, in Figure 1.1, K is the number of targets, which is defined by the output transcription file given by the ASR, which in our case is 4121 senones. The dimensionality of the bottleneck layer was fixed to bnf = 80; all these are typical values used in previous reports [47, 48].

BNF 2: The second approach combines information from standard MFCC, subband and residual information. The hypothesis in this case is that while the MFCCs contain useful phonetic information together with speaker de-pendent information important for normal speech related tasks, limited band and residual log-filterbank outputs contain mostly information re-lated to speaker identity that will be less dependent on the speech modal-ity, that is, whether it is normal or dysarthric speech. Hence, the feature vectors used as input to DNN are complementary to each other, and we ex-pect the resulting feature vector in the bottleneck layer to be more informa-tive than the BNF 1 feature set described above.

This is motivated by previous research exploring the role and relevance of dif-ferent frequency subbands for SR. As an example, for narrow-band speech sig-nals, the work in [49] showed that the 1.5–3.4 kHz frequency subband contains more discriminative information than the lower 0.3–1.5 kHz frequency sub-band, except for nasals. For wide-band speech signals, on the other hand, in [50] it was shown that the frequency subband 4–8 kHz provides performance similar to that obtained with the frequency subband 0–4 kHz, thus suggesting the presence of relevant speaker-discriminative information beyond 4 kHz. In a different study, it was shown that for text-dependent SID, higher-frequency channels were more relevant for SR than those located at lower frequencies [51]. It was reported that the lowest identification rates were associated with channels containing information of first and second formants, and that there was a high negative impact in performance when removing channels contain-ing information from the frequency band between 5 kHz and 8 kHz [51]. Regarding the information carried out by features extracted from residuals, in the past, these have been shown to contain important speaker-dependent infor-mation useful for SR tasks [52–54]. This is relevant for the task at hand because by removing the influence of the vocal tract, then it is possible to reduce the influence of atypical movement of articulators during speech production of dys-artric speakers and expected to carry complementary information that can be fused with other features [54, 55].

By using these insights, and taking into account our specific needs, we propose to train a deep neural network for BNF extraction using the same architecture as described for BNF 1 and using as input the following concatenated features: i) Thirteen MFCC, these features are aimed at the original task to train the DNN, subphonetic units classification. ii) Twenty-seven log Mel-scale filterbank outputs, the triangular filters are spaced between 1.2kHz and 8kHz. This range of frequencies was set after careful tuning using MFCC features only for normal speech and finding that the performance levels after dropping this acoustic band remained unaffected. Finally, a iii) 27 log Mel-scale filterbank outputs, extracted from the linear prediction residual. The input to the DNN is a time context of 13 frames, which was defined after an exploratory analysis. With this, the input to the DNN is a $d = 871$-dimensional vector $((13 + 27 + 27) \times 13)$. The bottleneck layer was also set in the third layer as depicted by Figure 1.1. The DNNs were trained using Theano [56] for both BNF 1 and BNF 2.

1.2.2 Utterance-based features

The frame-based representation produces a variable length sequence of feature vectors per speech recording. This representation, however, can be subject of further postprocessing that can be a suitable change (simplification or enrichment) of the representation. For example, by feature reduction, relations or primitives describing objects or some nonlinear transformation of the features can be done to enhance the class or cluster descriptions. Within the SR field, the well-known "supervector" is a fixed-length representation that can dramatically improve the performance of the classical GMM-MFCC based approach [6]. It is also possible to use a simpler approach in order to obtain a fixed-length feature vector from a frame-based representation, for example, statistic measures from individual time-varying variables (mean, variance, range, min, max) or to consider pairwise relations between variables (covariance or correlation). This approach allows to model the frame-based feature representation in terms of these measures and also to simplify and reduce the computational burden involved in more elaborated approaches such as i-vector extraction.

1.2.2.1 i-Vectors

The i-vectors extraction technique was proposed to map a variable length frame-based representation of an input speech recording, to a small-dimensional feature vector while retaining most relevant speaker information [17]. This approach models short-time-based features, such as MFCC, using a universal background model (UBM), typically a large GMM, where for parameter estimation high variability among speakers is required to guarantee a representative sample of the universe of speakers that are expected during the testing stage. The use of GMMs for SR is motivated by their capability to model arbitrary densities, and the individual components of a model are interpreted as broad acoustic classes [6]. Let D_f be the dimension of the *frame*-based features, N_c the number of components of the UBM and D_u the dimension of the *utterance*-based features (i.e., i-vectors); mathematically, a given speech recording can be represented by a GMM supervector M of dimension $(N_c{}^*D_f) \times 1$, which is assumed to be generated as follows:

$$M = m + Tx \tag{1.4}$$

where m of dimension $(N_c{}^*D_f) \times 1$ is the speaker-independent supervector (usually it can be taken to be the UBM supervector), T of dimension $(N_c{}^*D_f) \times D_u$ is a rectangular matrix of low rank, its column vectors spend the total variability space; x of dimension $D_u \times 1$ is a standard Gaussian-distributed latent vector and its maximum a posteriori point estimate provides the factors that best describe the utterance-dependent mean offset or what we simply name the i-vector. A maximum likelihood estimation procedure (EM-like algorithm) using a large training set of data is needed to estimate the total variability matrix T [17, 57]. Figure 1.2 depicts a diagram with the basic steps for i-vector computation.

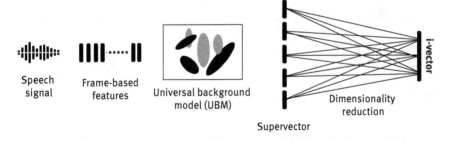

Speech signal

Frame-based features

Universal background model (UBM)

Supervector

Dimensionality reduction

i-vector

Figure 1.2: Basic idea of the i-vector features extraction from a given speech signal.

The challenge during parameter estimation is to learn as much variability as possible in order to properly model discriminative information among speakers. For the task at hand, this becomes an even more challenging problem, as to the best of our knowledge, no large-scale corpus with a large amount of speakers and different recording sessions is available for dysarthric speech. Then, it is expected that the feature sets used for i-vector extraction to be sufficiently informative, extracting highly discriminative speaker-dependent information independent of the speech modality.

1.2.2.2 Covariance features (CFs)

As previously mentioned, CFs are a simple representation of data using the sample covariance matrix. CFs were first used as a region descriptor for the problems of object detection and texture classification from images [24] and were later extended to dynamic data, such as video [25, 26] and audio [27]. The CF representation relays on the pairwise covariance relationship between the dimensions of the low-level region representations to model the distribution of the patterns presented in the feature space. Figure 1.3 illustrates the process of extracting covariance-based features from a speech signal. For a given input speech signal, we first extract a sequence of short-time features. The sample covariance matrix is then computed for such a sequence and the CF vector is obtained by vectorizing the upper (or lower) triangular part of the covariance matrix. Note that the dimension D_u of the resulting CF vector is given as follows:

$$D_u = \frac{D_f \times (D_f + 1)}{2}$$

where D_f is the dimension of the frame-based features.

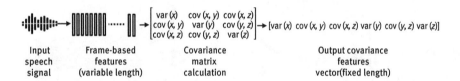

| Input speech signal | Frame-based features (variable length) | Covariance matrix calculation | Output covariance features vector(fixed length) |

Figure 1.3: An overview of the covariance-based feature extraction from a given speech signal.

The extraction of these features is rather simple as it does not need any extra background data,[1] contrary to the case of i-vectors or BNF features. These advantages make the covariance-based representation suitable for many applications dealing with limited data or limited computational resources.

---

[1] In the case of needing additional postprocessing, such as dimensionality reduction, additional background data would be required.

1.3 Experimental setup

1.3.1 Databases

Publicly available speech corpora containing normal and dysarthric speech are not common, and this is a challenging problem hampering research in this field. In this section we describe the publicly available speech databases used the experiments herein for SID purposes, and are the same databases used in [28], which will allow us to perform a direct comparison.

1.3.1.1 TORGO

This database contains approximately 23 h of North American English speech data from 15 speakers (9 males and 6 females) [58]. Eight speakers (5 males and 3 females) suffering from cerebral palsy or amyotrophic lateral sclerosis, which are two of the most prevalent dysarthria causes. The remaining seven speakers (4 males and 3 females) do not suffer from any form of speech disorder. A subset of approximately 8 h of TORGO database speech is available for free.[2] This subset contains a total of 9412 recordings (3177 dysarthric and 6235 controlled speech recordings) coming from all 15 speakers of the entire database. Each audio file is a RIFF (little-endian) WAVE format (Microsoft PCM, 16 bits, mono 16 kHz). In our experiments we used this subset of data including both dysarthric and controlled types of speech.

Various stimuli are included across all TORGO data, which can be regrouped into the following four categories: nonwords, short words, restricted sentences and unrestricted sentences.

1.3.1.2 NEMOURS

This database is a collection of 814 recordings (approximately one hour) from 11 male speakers with different dysarthria degrees [59]. Each speaker produce 74 short nonsense sentences and two connected-speech paragraphs. All the nonsense sentences have the same form: The N_0 is Ving the N_1, where N_0 and N_1 are unique monosyllabic nouns and **V** is a monosyllabic verb. Each

---

2 http://www.cs.toronto.edu/~complingweb/data/TORGO/torgo.html

nonsense sentence was saved in a wave file in standard RIFF format using a 16 kHz sampling rate at 16 bits precision.

Moreover, a 12th nondysarthric speaker is recruited to repeat all the nonsense sentences spoken by the 11 dysarthric speakers. The controlled speech part of NEMOURS database will not be used in this work since the task is SID and all the speech was produced by a single speaker.

1.3.1.3 TIMIT

This database is largely known in speech-processing-related fields. TIMIT contains broadband recordings of 630 speakers of eight major dialects of American English, each reading 10 phonetically rich sentences. It comprises 6300 speech recordings (approximately 5 h), recorded using 16 bits precision at 16 kHz. Average duration for each speech recordings is 3 s [60]. Even though the TIMIT corpus of read speech has been designed to provide speech data for acoustic–phonetic studies and for the development and evaluation of ASR systems, given the relatively large number of speakers, it is reasonable to be used also for SR applications.

1.3.1.4 LibriSpeech

This is a large-scale corpus of read English speech, and contains approximately 1000 h of speech derived from read audiobooks from the LibriVox project [47]. The speech is recorded using a sampling rate of 16 kHz using 16 bits precision. The data has been carefully segmented and aligned, which make it a suitable database for training ASR systems. In total, training data corresponds to 400 h of continuous speech. The aim of including this dataset is solely to train the BNF extractor system, which comprises two stages: first training an ASR system that generates the target labels or "senones," and next, training the bottleneck neural network using as input acoustic features extracted from the training speech recordings and as targets the senones generated by the ASR system as described in Section 2.1.4.

1.2 Classification

Speaker identification is the task of deciding, given a speech sample, who among a set of speakers said it. This is an N-class problem (given N speakers). In this work we are not interested in investigating the impact of the classifier on the final identification results. Therefore, we adopt a general-purpose classification model based on an SVM.

The SVM is a binary classifier that models the decision boundary between two classes as a separating hyperplane. By using labeled training vectors, the SVM optimizer finds a separating hyperplane that maximizes the separation between these two classes [61]. The discriminant function is given by

$$f(x) = \sum_{j=1}^{N} a_j c_j K(x, x_j) + b, \tag{1.5}$$

where $c_j \in \{+1, -1\}$ are the labels for the training vectors. The kernel function $K(\cdot, \cdot)$ is constrained to have certain properties (the Mercer condition). The support vectors x_j, their weights a_j and the bias term b are determined during training [61]. The decision score is calculated by comparing the test vector with the SVM discriminant function established during training, and a decision is made based on thresholding ("hard classification") [61]. Later, the SVM was effectively extended to perform the multiclass problems as well [62]. For the experiments herein, the open-source LIBSVM library [63] was used. Regarding the kernel function, after pilot experiments, it was found that the Gaussian kernel was optimal for the i-vector representation, whereas the polynomial kernel was optimal for the CFs.

1.3.3 Evaluation protocol

As previously mentioned, in this work we aim to explore the more recent advances in the SR field such as i-vectors, BNFs and utterance-level feature vectors based on covariance descriptors, for SID using dysarthric speech. Consequently, for the experiments herein, we adopted the following simple evaluation protocol:

1. Combine two well-known databases containing dysarthric speech (i.e., TORGO and NEMOURS) to be used as validation set.
2. Use control speakers (nondysarthric speakers) from the TORGO database as the validation set of the control speech (healthy speech).
3. Data from each validation set, dysarthric and control speech, was separated in two disjoint sets, train and test, with 70 and 30% of data, respectively.
4. For classification purposes, two multiclass SVMs were trained: the first one using only dysarthric speech as training data and a second one uses data from both dysarthric and healthy speakers in order to evaluate the effects of using mixed data for parameter estimation. All SVM hyperparameters

(kernel type, regularization constant c, kernel parameters: γ or the polynomial degree d) were tuned on the training set using a three-fold cross-validation procedure.

5. Finally, accuracy or identification rate (%) is reported as a performance measure for the two systems, using either only dysarthric speech or combined with control speech (healthy speakers).

As a benchmark, we rely on the results reported in [28], which, to our knowledge, is the only work published on dysarthric SR. In [28], both the TORGO and NEMOURS were used and the same 70/30 data split was utilized. The main difference is that the benchmark system relied on a subset of the two datasets (totaling 1330 files), thus removing files of very short duration in order to keep all files of similar duration. The benchmark system achieved, in this simpler task [64], an accuracy of 97.2%. Herein, we use this accuracy level to gauge our system performance.

1.4 Experimental results

In order to characterize the baseline system, we start by comparing performance of i-vectors and CFs on top of the standard MFCC feature vectors, and also by comparing the effects of using CMN or CMVN as normalization techniques. Results are summarized in Table 1.1. As can be seen, only for i-vectors, both CMN or CMVN negatively affect performance of the classification system in the two evaluation scenarios, that is, evaluating only with dysarthric speech or evaluating with combined dysarthric and control speech. By analyzing only the results for dysarthric speech, without using normalization the classification rate is 98.72%; this performance decreases to 95.56 and 92.33% using CMN and CMVN, respectively. For CFs, on the other hand, using a normalization based on CMVN seems to be optimal, as it helps to increase the performance from 86.84 to 97.29%. With this configuration, the CF representation has a competitive performance when compared with i-vectors. Furthermore, the fact that it is CMVN and not only CMN normalization the technique that has a positive effect on the CF representation suggests that the variance of short-term features introduce some undesirable effects in the high-level CF representation. Several factors could be the source of these undesirable effects, such as the short duration or the variable length of speech recordings.

In Table 1.1, the column labeled *Dysarthric and controlled speech* shows the accuracy achieved when using combined data to evaluate the classification system. These results help us to assess the capacity of the system to handle inputs

Table 1.1: Cross-validation (CV three-folds) and test set speaker identification results, as measured by the identification rate (%), of the i-vector and CF-based system built on top of MFCC frame-based features.

| | Normalization | Hyperparameters for test | | | | Dysarthric speech | | Dysarthric and controlled speech | |
		D_f	D_u	N_c	SVM	CV three-folds	Test	CV three-folds	Test
i-Vectors	CMN	60	150	512	Rbf, $c=5$, $\gamma=1$	95.41	95.56	92.03	92.86
	CMVN	60	150	512	Rbf, $c=5$, $\gamma=1$	92.55	92.33	88.31	89.99
	No norm.	60	200	512	Rbf, $c=10$, $\gamma=1$	98.42	**98.72**	97.43	97.53
CF	CMN	60	1830	–	Polyn., $d=2$, $c=10$, $\gamma=1$	84.40	86.84	79.14	82.15
	CMVN	60	1830	–	Polyn., $d=2$, $c=0.1$, $\gamma=8$	97.82	97.29	96.80	97.38
	No norm.	60	1830	–	Polyn., $d=2$, $c=10$, $\gamma=1$	84.40	86.84	79.14	82.15

independent of the speech modality, which in our case can be control or dysarthric speech. It can be seen that as a general rule, when using a speech-type independent system, lower performance is achieved, and losses have a relation ~1 to ~4% relative to the speech-type dependent system. There is one exception, however, that is when combining CF features with CMVN, in which case the speech-type independent system achieves a slightly better performance than its counterpart (97.38% vs 97.29%).

Table 1.2, in turn, reports similar experiments but using MS-based features as low-level features to compute i-vectors and the CF feature representation. As can be seen, the best performance achieved with this feature set is lower than the one obtained with MFCC (96.91% vs 98.72%). However, attained results are still competitive and represent high recognition rates, thus showing the discriminative capabilities of this feature representation. In particular, when using CF as postprocessing technique we can see that this approach is less sensitive to feature normalization methods, as the three normalization cases yield similar performances (see Figure 1.4). The MS feature extraction approach uses a rather long time context combined with PCA for dimensionality reduction; this process can actually be reducing linear channel effects, which is the whole purpose of techniques such as CMN and CMVN. Table 1.3, on the other hand, reports results by using the information contained in the modulation components in a slightly different way, that is, combining the instantaneous frequency and the Hilbert envelope to compute the WIF features (see Section 2.1.2). When comparing results obtained with i-vector versus CF feature representation, we can see that this feature set is more suitable for i-vector computation. Furthermore, when compared with MFCC, the WIF achieve a similar performance (98.42%) and also without any feature normalization process.

These two feature representations (i.e., MS and WIF) focus attention on slow varying information, which according to the results presented herein can model information about speaking rate and other speaker specific attributes. And together, results obtained from these two feature representations suggest that the information present in the slowly varying envelope of the bandpass signals is highly discriminative as has been shown before, and new feature extraction methods oriented to pathological speech need to be explored. However, for the task at hand, the MF is better posed to characterize the rate of change of long-term speech temporal envelopes [30], which can represent a problem in our experimental setup, where rather short-duration utterances are used (around one second in many cases). We hypothesize this is the reason the classical MFCC feature vectors perform better in this scenario. Auditory-inspired amplitude modulation features, however, can extract complementary information to MFCC and, according to our experiments, are more robust to normalization techniques, and

Table 1.2: Cross-validation (CV three-folds) and test set speaker identification results, as measured by the identification rate (%), of the i-vector and CF-based system built on top of MS frame-based features.

	Normalization	D_f	D_u	N_c	SVM	Dysarthric speech		Dysarthric and controlled speech	
						CV three-folds	Test	CV three-folds	Test
i-Vectors	cmn	40	400	256	Rbf, $c=10$, $\gamma=0.5$	91.92	91.72	89.70	89.14
	cmvn	40	400	128	Rbf, $c=5$, $\gamma=1$	92.51	93.30	88.31	90.86
	No norm.	40	400	128	Rbf, $c=10$, $\gamma=0.5$	96.69	**96.91**	94.83	94.19
CF	cmn	40	820	–	Polyn., $d=2$, $c=10$, $\gamma=8$	95.18	95.63	93.73	94.57
	cmvn	40	820	–	Polyn., $d=2$, $c=0.1$, $\gamma=4$	96.65	**96.91**	95.86	**96.33**
	No norm.	40	820	–	Polyn., $d=2$, $c=10$, $\gamma=8$	95.18	95.63	93.76	94.57

Hyperparameters for test

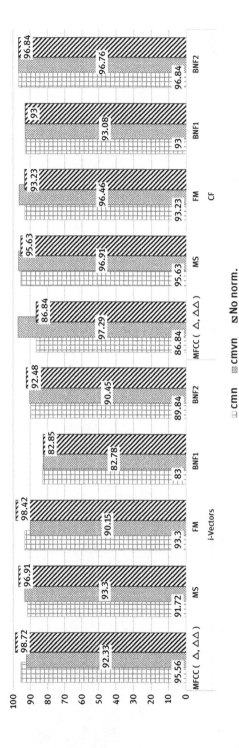

Figure 1.4: The impact of the short-term feature normalization methods (i.e., cmn, cmvn and no normalization) on the speaker identification rates (%) for all the combinations of the frame- and the utterance-based features. These results are taken from the "Test" columns of Tables 1.1, 1.2, 1.3 and 1.4.

Table 1.3: Cross-validation (CV three-folds) and test set speaker identification results, as measured by the identification rate (%), of the i-vector and CF-based system built on top of FM-WIF frame-based features.

| | Normalization | Hyperparameters for test | | | | Dysarthric speech | | Dysarthric and controlled speech | |
		D_f	D_u	N_c	SVM	CV three-folds	Test	CV three-folds	Test
i-Vectors	cmn	27	400	128	Rbf, $c=15$, $\gamma=0.5$	92.67	93.30	91.14	92.07
	cmvn	27	200	256	Rbf, $c=5$, $\gamma=0.5$	89.44	90.15	88.10	89.17
	No norm.	27	200	256	Rbf, $c=5$, $\gamma=1$	98.23	**98.42**	97.31	97.38
CF	cmn	27	378	–	Polyn., $d=2$, $c=10$, $\gamma=8$	92.48	93.23	92.51	93.92
	cmvn	27	378	–	Polyn., $d=2$, $c=0.1$, $\gamma=8$	96.54	96.46	95.96	95.89
	No norm.	27	378	–	Polyn., $d=2$, $c=10$, $\gamma=8$	92.48	93.23	92.51	93.92

Table 1.4: Cross-validation (CV three-folds) and test set speaker identification results, as measured by the identification rate (%), of the i-vector and CF-based system built on top of two versions of BNF frame-based features.

		MFCC configuration		Hyperparameters for test			Dysarthric speech		Dysarthric and controlled speech	
	BNF version	Normalization	D_f	D_u	N_c	SVM	CV three-folds	Test	CV three-folds	Test
i-Vectors	BNF 1	cmn	80	300	256	Rbf, $c=5$, $\gamma=1$	81.24	83.00	70.06	71.91
		cmvn	80	200	256	Rbf, $c=5$, $\gamma=1$	81.66	82.78	70.82	72.68
		No norm.	80	300	256	Rbf, $c=5$, $\gamma=0.5$	82.37	82.85	70.44	70.80
	BNF 2	cmn	80	200	256	Rbf, $c=5$, $\gamma=1$	87.78	89.84	81.51	83.97
		cmvn	80	200	256	Rbf, $c=5$, $\gamma=1$	87.52	90.45	81.31	83.27
		No norm.	80	200	256	Rbf, $c=5$, $\gamma=1$	92.07	92.48	86.47	87.26
CF	BNF 1	cmn	80	3240	–	Polyn., $d=2$, $c=1$, $\gamma=2$	92.78	93.00	88.22	88.82
		cmvn	80	3240	–	Polyn., $d=2$, $c=1$, $\gamma=2$	92.78	93.08	88.23	88.84
		No norm.	80	3240	–	Polyn., $d=2$, $c=1$, $\gamma=2$	92.78	93.00	88.22	88.82
	BNF 2	cmn	80	3240	–	Polyn., $d=2$, $c=5$, $\gamma=1$	95.15	**96.84**	93.26	94.45
		cmvn	80	3240	–	Polyn., $d=2$, $c=0.5$, $\gamma=4$	95.26	96.76	92.90	94.21
		No norm.	80	3240	–	Polyn., $d=2$, $c=1$, $\gamma=2$	95.11	**96.84**	93.23	94.48

thus should represent an advantage when using fusion schemes to combine strengths of different feature representations.

Finally, Table 1.4 reports results for the bottleneck (BNF)-based features. As can be seen, contrary to the effect observed with WIF and MFCC, both BNF versions (i.e., BNF 1 and BNF 2) are more suitable for CF representation than for i-vectors. According to the results, BNF 2 exhibits better performance when compared to BNF 1. It can be seen that either using i-vectors extraction or CF as postprocessing technique, information contained in BNF 2 is more discriminative. Moreover, similar to what was observed with MS features, BNF 2 is less sensitive to feature normalization methods when using CF (see Figure 1.4). While using MFCC and WIF features resulted in superior accuracy relative to BNF-based approaches, the specific scenario we are evaluating these features can affect their final performance. For example, short-duration utterances limits the phonetic variability present in the testing utterances; this can affect BNF-based approaches while it at the same time benefit standard MFCC features [65, 66].

Furthermore, while for standard short-time based features, such as MFCC, MS or WIF (which do not require a training stage), we suggest to explore fusion schemes in order to combine strengths of individual features. In the case of BNF features we need to pay attention to the effect that has changing the input feature vectors to the DNN. According to our results, by using acoustic and/or perceptual analyses to motivate the concatenated feature vectors used as input to train the DNN model, the resulting BNF feature vector had improved accuracy results. Hence, prior to explore fusion techniques, it is necessary to explore alternate configurations and inputs to better benefit from the capabilities of DNN-based feature extraction in these atypical scenarios. This problem will be explored in a future work.

1.5 Conclusion

This chapter has addressed the issue of SID based on dysarthric speech. Several state-of-the-art processing techniques commonly applied in healthy speech SID were explored to gauge their effectiveness under atypical scenarios. More specifically, four different low-level feature extraction methods were compared, using two different postprocessing techniques: i-vector extraction and covariance-features representation, and different feature normalization effects were explored. Overall, the best identification result (98.7%) was achieved using i-vectors extracted from MFCC features without feature normalization. These results compare favorably with those previously reported in the literature, where the best accuracy reported to date was of 97.2%, but on a subset of the two datasets used herein.

Acknowledgment: The authors acknowledge funding from FRQNT and the Administrative Department of Science, Technology and Innovation of Colombia (COLCIENCIAS).

References

[1]   Levi, S. V., and Pisoni, D. B. Indexical and linguistic channels in speech perception: Some effects of voiceovers on advertising outcomes, Psycholinguistic phenomena in marketing communications, 203–219, 2007.

[2]   O'Shaughnessy, D. Speech communications – human and machine (2. ed.). 1em plus 0.5em minus 0.4em IEEE, 2000.

[3]   Benzeghiba, M., Mori, R. D., Deroo, O., Dupont, S., Erbes, T., Jouvet, D., Fissore, L., Laface, P., Mertins, A., Ris, C., Rose, R., Tyagi, V., and Wellekens, C. Automatic speech recognition and speech variability: A review. Speech Communication, 49(10), 763–786, 2007.

[4]   O'Shaughnessy, D. Invited paper: Automatic speech recognition: History, methods and challenges. Pattern Recognition, 41(10), 2965–2979, 2008.

[5]   Yu, D., and Deng, L. Automatic Speech Recognition: A Deep Learning Approach, 1em plus 0.5em minus 0.4em Springer Publishing Company, Incorporated, 2014.

[6]   Kinnunen, T., and Li, H. An overview of text-independent speaker recognition: From features to supervectors. Speech Communication, 52(1), 12–40, January 2010.

[7]   Torre, P., and Barlow, J. A. Age-related changes in acoustic characteristics of adult speech. Journal of communication disorders, 42(5), 324–333, 2009.

[8]   Kain, A. B., Hosom, J.-P., Niu, X., van Santen, J. P., Fried-Oken, M., and Staehely, J. Improving the intelligibility of dysarthric speech. Speech communication, 49(9), 743–759, 2007.

[9]   Doyle, P. C., Leeper, H. A., Kotler, A.-L., Thomas-Stonell, N. et al. Dysarthric speech: A comparison of computerized speech recognition and listener intelligility. Journal of Rehabilitation Research and Development, 34(3), 309, 1997.

[10] Kent, R. D., Weismer, G., Kent, J. F., Vorperian, H. K., and Duffy, J. R. Acoustic studies of dysarthric speech: Methods, progress, and potential. Journal of communication disorders, 32(3), 141–186, 1999.

[11] Duffy, J. R. Motor speech disorders: Substrates, differential diagnosis, and management, 1em plus 0.5em minus 0.4em, Elsevier Health Sciences, 2013.

[12] Brown, A. Social Aspects of Communication in Parkinson's Disease, 1em plus 0.5em minus 0.4em, De Montfort University, 2013.

[13] Falk, T., and Chan, W.-Y. Temporal dynamics for blind measurement of room acoustical parameters. Instrumentation and Measurement, IEEE Transactions on, 59(4), 978–989, April 2010.

[14] Rudzicz, F. Adjusting dysarthric speech signals to be more intelligible. Computer Speech & Language, 27(6), 1163–1177, 2013.

[15] Rajeswari, N., and Chandrakala, S. Generative model-driven feature learning for dysarthric speech recognition. Biocybernetics and Biomedical Engineering, 36(4), 553–561, 2016.

[16] Unar, J., Chaw Seng, W., and Abbasi, A. A review of biometric technology along with trends and prospects. Pattern Recognition, 47(8), 2673–2688, 2014.

[17] Dehak, N., Kenny, P., Dehak, R., Dumouchel, P., and Ouellet, P. Front end factor analysis for speaker verification. IEEE Transactions on Audio, Speech and Language Processing, 2010.

[18] Dehak, N., Torres-Carrasquillo, P. A., Reynolds, D., and Dehak, R. "Language Recognition via I-Vectors and Dimensionality Reduction," in INTERSPEECH 2011, Florence, Italy, Aug. 2011, pp. 857–860.

[19] Martinez, D., Burget, L., Ferrer, L., and Scheffer, N. "ivector-based prosodic system for language identification," in Acoustics, Speech and Signal Processing (ICASSP), 2012 IEEE International Conference on, March 2012, pp. 4861–4864.

[20] Soufifar, M., Kockmann, M., s Burget, L., Plchot, O., Glembek, O., and Svendsen, T. ivector approach to phonotactic language recognition, INTERSPEECH, 2011.

[21] Martinez, D., Lleida, E., Ortega, A., and Miguel, A. "Prosodic features and formant modeling for an ivector-based language recognition system," in Acoustics, Speech and Signal Processing (ICASSP), 2013 IEEE International Conference on, May 2013, pp. 6847–6851.

[22] Saon, G., Soltau, H., Nahamoo, D., and Picheny, M. "Speaker adaptation of neural network acoustic models using i-vectors,". Automatic Speech Recognition and Understanding (ASRU), 2013 IEEE Workshop on, Dec 2013, 55–59.

[23] Martinez, D., Green, P., and Christensen, H. "Dysarthria intelligibility assessment in a factor analysis total variability space," in Proceedings of Interspeech, 2013.

[24] Tuzel, O., Porikli, F., and Meer, P. "Region covariance: A fast descriptor for detection and classification," in European conference on computer vision. 1em plus 0.5em minus 0.4em Springer, 2006, pp. 589–600.

[25] Porikli, F., Tuzel, O., and Meer, P. "Covariance tracking using model update based on lie algebra," in Computer Vision and Pattern Recognition, 2006 IEEE Computer Society Conference on, vol. 1, June 2006, pp. 728–735.

[26] Tuzel, O., Porikli, F., and Meer, P., "Human detection via classification on Riemannian manifolds," in Computer Vision and Pattern Recognition, 2007. CVPR'07. IEEE Conference on, June 2007, pp. 1–8.

[27] Ye, C., Liu, J., Chen, C., Song, M., and Bu, J. Advances in Multimedia Information Processing – PCM 2008: 9th Pacific Rim Conference on Multimedia, Tainan, Taiwan, December 9–13, 2008. Proceedings, 1em plus 0.5em minus 0.4em Berlin, Heidelberg, Springer Berlin Heidelberg, 2008, ch. Speech Emotion Classification on a Riemannian Manifold, 61–69.

[28] Kadi, K. L., Selouani, S. A., Boudraa, B., and Boudraa, M. Fully automated speaker identification and intelligibility assessment in dysarthria disease using auditory knowledge. Biocybernetics and Biomedical Engineering, 36(1), 233–247, 2016.

[29] Sarria-Paja, M., and Falk, T. H. Automated dysarthria severity classification for improved objective intelligibility assessment of spastic dysarthric speech, Proc. INTERSPEECH, 2012, 62–65.

[30] Falk, T. H., Chan, W.-Y., and Shein, F. Characterization of atypical vocal source excitation, temporal dynamics and prosody for objective measurement of dysarthric word intelligibility. Speech Communication, 54(5), 622–631, 2012.

[31] Sadjadi, S., and Hansen, J. Assessment of single-channel speech enhancement techniques for speaker identification under mismatched conditions. Proc. In INTERSPEECH, 2138–2141, 2010.

[32] Grimaldi, M., and Cummins, F. Speaker identification using instantaneous frequencies. IEEE Transactions on Audio, Speech, and Language Processing, 16(6), 1097–1111, August 2008.

[33] Sarria-Paja, M., and Falk, T. H. Fusion of auditory inspired amplitude modulation spectrum and cepstral features for whispered and normal speech speaker verification. Computer Speech & Language, 45, 437–456, 2017.

[34] Richardson, F., Reynolds, D., and Dehak, N. "A unified deep neural network for speaker and language recognition," arXiv:1504.00923, 2015.

[35] Rabiner, L., and Schafer, R. Digital processing of speech signals, ser. Prentice-Hall signal processing series. 1em plus 0.5em minus 0.4em Englewood Cliffs, N.J. Prentice-Hall, 1978.

[36] Alam, M. J., Ouellet, P., Kenny, P., and O'Shaughnessy, D. "Comparative evaluation of feature normalization techniques for speaker verification," in International Conference on Nonlinear Speech Processing. 1em plus 0.5em minus 0.4em Springer, 2011, pp. 246–253.

[37] Falk, T., and Chan, W.-Y. Modulation spectral features for robust far-field speaker identification. IEEE Transactions on Audio, Speech, and Language Processing, 18(1), 90–100, January 2010.

[38] Sarria-Paja, M., and Falk, T. H. Whispered speech detection in noise using auditory-inspired modulation spectrum features. IEEE Signal Processing Letters, 20(8), 783–786, 2013.

[39] Xiang, J., Poeppel, D., and Simon, J. "Physiological evidence for auditory modulation filterbanks: Cortical responses to concurrent modulations,". JASA-EL, 133(1), EL7–El12, 2013.

[40] Kinnunen, T., Lee, K., and Li, H. "Dimension reduction of the modulation spectrogram for speaker verification," in Proc. The Speaker and Language Recognition Workshop (Odyssey 2008), 2008.

[41] McLaren, M., Ferrer, L., and Lawson, A. Exploring the role of phonetic bottleneck features for speaker and language recognition. Proc. ICASPP, 5575–5579, March 2016.

[42] Reddy, S. R.,M. R., and Umesh, S. "Improved acoustic modeling for automatic dysarthric speech recognition," in National Conference on Communications (NCC), Feb 2015, pp. 1–6.

[43] Kim, H., Hasegawa-Johnson, M., and Perlman, A. Vowel contrast and speech intelligibility in dysarthria. Folia Phoniatrica et Logopaedica, 63(4), 187–194, 2010.

[44] Sharifzadeh, H., McLoughlin, I., and Russell, M. A comprehensive vowel space for whispered speech. Journal of Voice, 26(2), 49–56, March 2012.

[45] Sarria-Paja, M., and Falk, T. "Strategies to enhance whispered speech speaker verification: A comparative analysis,". Journal of the Canadian Acoustical Association, 43(4), 31–45, 2015.

[46] Povey, D., Ghoshal, A., Boulianne, G., Burget, L., Glembek, O., Goel, N., Hannemann, M., Motlicek, P., Qian, Y., Schwarz, P., Silovsky, J., Stemmer, G., and Vesely, K. "The kaldi

speech recognition toolkit," in IEEE Workshop on Automatic Speech Recognition and Understanding, December 2011.

[47] Panayotov, V., Chen, G., Povey, D., and Khudanpur, S. Librispeech: An asr corpus based on public domain audio books. Proc. ICASSP, April 2015, 5206–5210.

[48] Matejka, P., Glembek, O., Novotny, O., Plchot, O., GrÃ©zl, F., Burget, L., and Cernocky, J. Analysis of DNN approaches to speaker identification. Proc. ICASSP, March 2016, 5100–5104.

[49] Lei, H., and Lopez-Gonzalo, E. Mel, linear, and antimel frequency cepstral coefficients in broad phonetic regions for telephone speaker recognition, Proc. INTERSPEECH, 2009.

[50] Gallardo, L., Wagner, M., and Möller, S., "Advantages of wideband over narrowband channels for speaker verification employing mfccs and lfccs." in Proc. INTERSPEECH, 2014.

[51] Besacier, L., and Bonastre, J.-F. Subband approach for automatic speaker recognition: Optimal division of the frequency domain. Proc. AVBPA, 195–202, ser. Lecture Notes in Computer Science, March 1997.

[52] Drugman, T., and Dutoit, T. "On the potential of glottal signatures for speaker recognition," in Proc. INTERSPEECH. 1em plus 0.5em minus 0.4em ISCA, 2010, pp. 2106–2109.

[53] Chetouani, M., Faundez-Zanuy, M., Gas, B., and Zarader, J. Investigation on lp-residual representations for speaker identification,'. Pattern Recognition, 42(3), 487–494, 2009.

[54] Debadatta, P., and Prasanna, M. Processing of linear prediction residual in spectral and cepstral domains for speaker information. International Journal of Speech Technology, 18(3), 333–350, 2015.

[55] Sahidullah, M., Chakroborty, S., and Saha, G. "Improving performance of speaker identification system using complementary information fusion,". CoRR, abs/1105.2770, 2011.

[56] Theano Development Team "Theano: A Python framework for fast computation of mathematical expressions,", arXiv e-prints, abs/1605.02688, may, 2016.

[57] Kenny, P. A small footprint i-vector extractor, Odyssey, 2012, 1–6.

[58] Rudzicz, F., Namasivayam, A. K., and Wolff, T. The torgo database of acoustic and articulatory speech from speakers with dysarthria. Language Resources and Evaluation, 46(4), 523–541, 2012.

[59] Menendez-Pidal, X., Polikoff, J. B., Peters, S. M., Leonzio, J. E., and Bunnell, H. T. "The nemours database of dysarthric speech," in Spoken Language, 1996. ICSLP 96. Proceedings., Fourth International Conference on, vol. 3. 1em plus 0.5em minus 0.4em IEEE, 1996, pp. 1962–1965.

[60] Garofolo, J. S., and Consortium, L. D.et al. "TIMIT: acoustic-phonetic continuous speech corpus," 1993.

[61] Vapnik, V. N., and Vapnik, V. Statistical learning theory, 1em plus 0.5em minus 0.4em, Wiley New York, 1998, Vol. 1.

[62] Hsu, C.-W., and Lin, C.-J. A comparison of methods for multiclass support vector machines. IEEE transactions on Neural Networks, 13(2), 415–425, 2002.

[63] Chang, C.-C., and Lin, C.-J. LIBSVM: A library for support vector machines. ACM Transactions on Intelligent Systems and Technology, 2, 27:1–27:27, 2011, software available at. http://www.csie.ntu.edu.tw/cjlin/libsvm.

[64] Kenny, P., Stafylakis, T., Ouellet, P., Alam, M. J., and Dumouchel, P. "Plda for speaker verification with utterances of arbitrary duration," in Acoustics, Speech and Signal

Processing (ICASSP), 2013 IEEE International Conference on. 1em plus 0.5em minus 0.4em IEEE, 2013, pp. 7649–7653.

[65] Vogt, R. J., Lustri, C. J., and Sridharan, S., "Factor analysis modelling for speaker verification with short utterances," in Proc. The IEEE Odyssey Speaker and Language Recognition Workshop. 1em plus 0.5em minus 0.4em IEEE, 2008.

[66] Kanagasundaram, A., Vogt, R., Dean, D., Sridharan, S., and Mason, M. "I-vector based speaker recognition on short utterances,". Proc. INTERSPEECH, 2341–2344, 2011.

Anusha Prakash, M. Ramasubba Reddy and Hema A. Murthy

2 Enhancement of continuous dysarthric speech

Abstract: Dysarthria is a set of motor–speech disorders resulting due to neurological injuries. It affects the motor component of the motor–speech system. Disruption in muscular control makes the speech imperfect. As a result, dysarthric speech is not as comprehensible as normal speech. Most often, people with dysarthria have problems with communicating and this inhibits their social participation. Hence, for effective communication, it is extremely vital that we develop assistive speech technologies for people with dysarthria. The aim is to improve the naturalness and intelligibility of dysarthric speech while retaining the characteristics of the speaker. For this purpose, durational attributes across dysarthric and normal speech utterances are first studied. An automatic technique is developed to correct dysarthric speech to bring it closer to normal speech. The performance of this technique is compared with that of two other techniques available in the literature – a formant resynthesis technique and a hidden Markov model based adaptive speech synthesis technique. Subjective evaluations show a preference for dysarthric speech modified using the proposed approach over existing approaches.

Keywords: continuous dysarthric speech, Indian dysarthric speaker, durational modifications, dynamic time-warping

2.1 Introduction

Dysarthria refers to a group of neuromuscular speech disorders [1]. These speech disorders are a result of neurological injury. Dysarthria can be developmental (cerebral palsy) or acquired. The causes of acquired dysarthria include degenerative diseases (multiple sclerosis, Parkinson's disease), brain injuries and strokes affecting neuromuscular control. Based on the region of the nervous system that is affected, types of dysarthria are spastic, ataxic, flaccid, hyperkinetic, hypokinetic and mixed.

Anusha Prakash, M. Ramasubba Reddy, Hema A. Murthy, Indian Institute of Technology Madras, India

https://doi.org/10.1515/9781501501265-003

The word dysarthria, originating from *dys* and *arthrosis*, means difficult or imperfect articulation. Dysarthria affects any of the speech subsystems such as respiration, phonation, resonance, articulation and prosody. The physical production of speech is impaired. Disruption in muscular control makes the speech imperfect. The characteristics of dysarthric speech are poorly articulated phonemes, problems with speech rate, uneven speech volume, the voicing of unvoiced units, slurring, incorrect pitch trajectory and so on. The person with dysarthria may have additional problems of drooling, swallowing while speaking. As a result, dysarthric speech is not as comprehensible as normal speech.

Most often, people with dysarthria have problems with communicating and this inhibits their social interaction. This may lead to issues of low self-esteem and depression. It can also adversely affect the rehabilitation of people with dysarthria. Due to speech impairments and their associated problems, people with dysarthria are mostly sidelined from mainstream education, job sector and so on. Hence, it is imperative that assistive speech technologies are developed that enable people with dysarthria to communicate effectively.

The quality of continuous dysarthric speech needs to be improved to assist people with dysarthria communicate better. The objective of this work is to create aids for people with dysarthria such that given a dysarthric speech utterance, more natural and intelligible speech is produced while retaining the characteristics of the speaker. The focus is on continuous speech rather than isolated words (or subword units). Given the multiple causes and types of dysarthria, the challenge is in developing a generic technique that can enhance dysarthric speech across a wide range of types and levels of severity.

Durational attributes across dysarthric and normal speech utterances are first studied for enhancing dysarthric speech. An automatic technique is developed to correct dysarthric speech closer to normal speech. In other words, normal speech characteristics are transplanted onto dysarthric speech to improve the latter's quality while preserving the speaker's voice characteristics. The performance of this technique is compared with that of two other techniques available in the literature – a formant resynthesis technique and a hidden Markov model (HMM)-based adaptive speech synthesis technique.

The previous attempts to enhance dysarthric speech are first reviewed. The standard dysarthric speech databases – Nemours database [2] and TORGO database [3], and an Indian English dysarthric speech dataset used in the experiments are then detailed. The implementation of the formant resynthesis and the HMM-based adaptive speech synthesis techniques are described. The proposed modifications to dysarthric speech, including the durational analyses, are presented. To compare the performance of the proposed technique with available methods for enhancement, subjective evaluations are conducted. This work has been published by us in SLPAT (2016) [4] and has been extended to include experiments with TORGO database in the current work.

2.2 Related work

There have been several efforts in speech recognition [5, 6] and automatic correction/synthesis of dysarthric speech [7–9]. In Hosom et al. [7], dysarthric and normal phoneme feature vectors for each utterance are aligned using dynamic time warping (DTW), and then a transformation function is learnt to correct dysarthric speech. The word-level intelligibility of dysarthric speech utterances of speaker LL in Nemours database is improved from 67% to 87%. The drawback of this method is that labeling and segmentation of dysarthric and normal speech need to be manually verified and corrected. In Rudzicz [8, 10], dysarthric speech is improved by correcting pronunciation errors and by modifying the waveform in time and frequency. The author reports that this kind of modification does not improve the intelligibility of dysarthric speech. Moreover, to correct pronunciation errors, corresponding transcriptions are required.

In Yakcoub et al. [11], dysarthric speech is passed to an HMM-based speech recognizer and wrongly uttered units are corrected using a grafting technique. In Dhanalakshmi et al. [12], dysarthric speech is first recognized and then synthesized by speaker adaptation using HMM-based techniques. In Saranya et al. [13], poorly uttered phonemes in dysarthric speech are replaced by phonemes from normal speech. Discontinuities present in the contours of short-term energy, pitch and formant at concatenation points are addressed. In Yakcoub et al. [11] and Saranya et al. [13], the poorly uttered units in dysarthric speech need to be identified. Kain et al. [9, 14] explore a format resynthesis technique using transformed formants, smoothened energy and synthetic pitch contours. This has shown to improve the intelligibility of vowels in isolated words spoken by a dysarthric person.

To improve the intelligibility of dysarthric speech, two techniques available in the literature are implemented. These are baseline systems. The first technique is the formant resynthesis approach [9] with few modifications to suit the data used in the experiments. This method is explained in Section 4. The second technique is the HMM-based text-to-speech (TTS) synthesis approach by adapting to the dysarthric person's voice [12]. It is assumed that a 100% accurate recognition system is already available to transcribe dysarthric speech for synthesis. The HMM-based technique is discussed in Section 5.

2.3 Data used

Standard databases available for dysarthric speech in English are Nemours [2], TORGO [3] and Universal Access [15]. Nemours and TORGO databases contain recordings of complete sentences spoken by people with dysarthria and are used in this work. Universal Access database contains audiovisual isolated word recordings and is hence not suitable for our purpose. There is also a work published on the collection of Tamil dysarthric speech corpus [16]. At the time of this writing, this data is not publicly available.

Additionally, the speech of a native Indian having dysarthria is collected and used in the experiments. The motive for collecting the Indian English dataset is twofold:
- To make use of unstructured text, unlike the structured text present in Nemours database (explained in Section 3.1).
- To assess the performance of the techniques for dysarthric speech of a native Indian.

The Indian English dysarthric speech data is referred to as "IE" in this work.[1] Analysis and experiments were initially performed with Nemours database and IE dataset. TORGO database has been used only to validate the results of the proposed technique.

2.3.1 Nemours database

Nemours database [2] consists of dysarthric speech data of 11 North American male speakers. Speakers are classified based on the degree of severity of dysarthria: mild (BB, FB, LL, MH), moderate (JF, RK, RL) and severe (BK, BV, KS, SC). Corresponding to the recording of each dysarthric speaker, speech of a normal speaker is provided. Labels of normal speakers have a prefix "JP." For example, normal speaker corresponding to dysarthric speaker BB is JPBB.

The Nemours speech data consists of 74 nonsense sentences and two paragraphs for each speaker. Since phone-level transcriptions for the paragraphs are not provided, speech data of the paragraphs is excluded from the experiments. The sentences follow the same format: *The X is Y'ing the Z*, where X and Z are

---

1 The Indian English dysarthric speech data can be found at www.iitm.ac.in/donlab/website_files/resources/IEDysarthria.zip

selected from a set of 74 monosyllabic nouns with $X{\neq}Z$, and *Y'ing* is selected from a set of 37 bisyllabic verbs. An example of such a structured sentence is – *The bash is pairing the bath*. Phone-level transcriptions of each word are provided in terms of Arpabet labels [17].

Significant intra-utterance pauses are observed in the speech of dysarthric speakers BK, RK, RL and SC. For speaker RK, pauses within an utterance were already marked. However, intra-utterance pauses are not provided for speakers BK, RL and SC, and therefore pauses were manually marked for these speakers. Phonemic labeling is not provided for speech data of speaker KS. Hence, data of KS is excluded from the experiments.

Phone-level alignment is provided for dysarthric speech data while only word-level alignment is provided for normal speech data. The procedure to obtain phone-level alignment from word-level alignment for normal speech data is described.

2.3.1.1 Alignment of normal speech data at phone level

Normal speech data is aligned at the phone level using HMM-based approach. Phone transcriptions for each word and word-level boundaries are provided in the database. HMMs are used to model source and system parameters of monophones in the data. The source features are pitch ($\log f0$) values and the system parameters are mel frequency cepstral coefficients (MFCC). As speech is dynamic in nature, velocity and acceleration values of the features are also included. Embedded training of HMM parameters is restricted to the word boundary rather than performing it on the entire sentence. This is inspired by [18], in which better phone-level boundaries are obtained from embedded training within syllable boundaries.

The simplest approach to training HMMs is the flat start method, in which models are initialized with global mean and variance. However, to obtain better phone boundaries, HMMs built using Carnegie Mellon University (CMU) corpus [19] were used as initial monophone HMMs. For this purpose, around 1 hour of speech data of American speaker "rms" in CMU corpus was used.

2.3.2 TORGO database

TORGO database consists of articulatory and acoustic data of complete sentences, short words and nonwords [3]. Speech of 8 dysarthric speakers (3 females and 5 males) and 7 control subjects (3 females and 4 males) are available in the database. This database is used only to validate the performance of the proposed automatic technique (presented in Section 9) and is not used for implementing the previous techniques of correction.

The proposed technique uses only the audio files and does not consider the corresponding labels. Wave files recorded with the head-mounted microphone are used; if they are not available, then those recorded using array microphones are used. Only complete sentences are retained. Duplicate sentences are removed manually after listening to the wave files. Continuous speech of dysarthric and normal speakers are matched using the given text. Speech of male speakers with moderate to severe dysarthria (M01, M02, M04, M05) are used in the experiments. Speech of speaker MC03 is considered as the normal speech in the experiments.

2.3.3 Indian English dysarthric speech dataset

The process of text selection, speech recording and alignment of the Indian English dysarthric speech data is detailed.

2.3.3.1 Text selection

Seventy-three sentences were selected from CMU corpus [19] such that they ensured enough phone coverage. These sentences are unstructured. Phone transcriptions of words in the text were obtained from CMU pronunciation dictionary [20]. The phone transcriptions were manually checked and corrected to suit Indian English pronunciation. To account for pauses or silences, an additional label "pau" was added.

2.3.3.2 Speech recording

Speech of a mildly dysarthric Indian male suffering from cerebral palsy was recorded in a low-noise environment. The speech was sampled at 16 kHz, with 16 significant bits. The recording was performed over several sessions, with each session at most 30 min in duration. As per the convenience of the speaker, frequent breaks were given during the sessions. This was done to eliminate the effect of fatigue on the quality of speech. About 11 min of speech data was collected. Due to the unavailability of a speech pathologist, Frenchay dysarthria assessment (FDA) [21] was not performed.

2.3.3.3 Phone-level alignment

Before alignment, voice activity detection was used to remove long-silence regions (more than 100 ms) from the speech waveforms. About 11 min of data then reduced to 8.5 min. Alignment was performed semi-automatically. As speaker IE is a native Malayalam speaker, HMMs were trained from already available normal English speech data of an Indian (Malayalam) speaker "IEm" [22]. These monophone HMMs were used as initial HMMs to align dysarthric speech data at the phone level. Alignment was then manually inspected and corrected. Details of the IE dataset are given in Table 2.1.

Table 2.1: Details of Indian English dysarthric speech dataset.

Number of sentences	73
Number of unique words	369
Total duration	8.5 min

2.4 Formant resynthesis technique

Kain et al. [9] improve the intelligibility of dysarthric vowels in isolated words. These words are of type CVC type (C, consonant; V, vowel). In this work, a similar approach is adopted to improve the intelligibility of continuous dysarthric speech. Formant frequencies F1–F4, short-term energy and pitch values are extracted from dysarthric and normal speech. Window attributes of 25 ms length and 10 ms shift are used. Formant transformation from dysarthric space to normal space is only carried out in the vowel regions. Utterances have to be first aligned at the phone level as vowel boundaries and vowel identities are required. Formant values in the stable region of the vowels are then determined [9]. The stable region (or point) is the vowel region (or point) that is least affected by context. A four-dimensional feature vector is used for representing each instance of a vowel – F1stable, F2stable, F3stable and vowel duration.

In Kain et al. [9], Gaussian mixture model (GMM) parameters are trained by joint density estimation for formant transformation. This is effective when the data is phonetically balanced. The data used in this work suffer from data imbalance problem as there is variation in the frequency of individual vowels. To address this, a universal background model-GMM (UBM-GMM) [23] is trained and adapted to individual vowels of dysarthric and normal speech. The adaptation method used here is maximum a posteriori (MAP). An illustration of the UBM-GMM and its mean adapted model are shown in Figure 2.1. The procedure to obtain adapted models is described briefly:

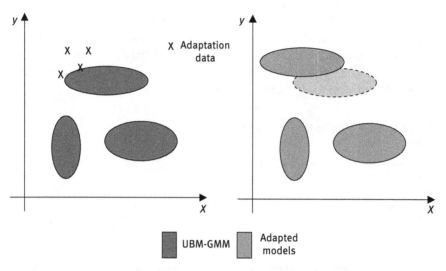

Figure 2.1: Example of UBM-GMM and mean-adapted UBM-GMM.

- Frames in vowel regions are represented by four-dimensional feature vectors. These dimensions are formants F1, F2, F3 and vowel duration. Irrespective of stable points, all the feature vectors are pooled together for all vowel instances across dysarthric and normal speech. A UBM-GMM is trained for the pooled feature vectors.
- All instances of a particular vowel in the adaptation data (dysarthric or normal speech) are represented by four-dimensional feature vectors (F1stable, F2stable, F3stable and vowel duration). Here, a single vowel region is represented by a single feature vector and not multiple framewise representations as in training the UBM-GMM.
- After adapting only the means of the UBM-GMM using the adaptation data, a codebook of means is obtained across dysarthric and normal speech. The codebook contains a total of $2 \times$ number_of_vowels entries.

For the dysarthric speech in Nemours database, corresponding normal speech is used for generating the codebook. Speech of speaker "ksp" from CMU corpus [19] is used as the normal speech for the Indian dysarthric speech.

For the experiments, dysarthric speech data is divided into train (80%) and test data (20%). The train data is used for building the adapted models. The codebook size of the UBM-GMM is 64.

A flowchart illustrating the resynthesis of dysarthric speech is shown in Figure 2.2. From the test data, energy and pitch contours are extracted and smoothened. For smoothening, the contours are passed through a median filter of order 3 and then low passed using a Hanning window. In Kain et al. [9], a synthetic F0 contour is used for resynthesis instead of the smoothened pitch contour. Using the alignment information, every vowel in the test utterance is represented by a four-dimensional feature vector (stable F1–F3, vowel duration). Feature vectors of the dysarthric vowels are replaced by the means of their normal counterpart using the codebook of means. The entire vowel is represented by the transformed or replaced stable point formants. Hence, the same stable point formant value is repeated across the frames of the vowel. Speech is synthesized using the transformed formant contours, smoothened energy and pitch contours [24]. Non-vowel regions in the resynthesized dysarthric speech are replaced by the original dysarthric speech to produce the final modified dysarthric speech.

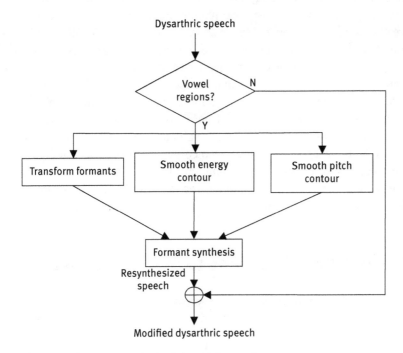

Figure 2.2: Formant resynthesis of dysarthric speech.

2.5 HMM-based TTS using adaptation

In Dhanalakshmi et al. (2015), an HMM-based TTS synthesizer (HTS) is developed by adapting to the dysarthric person's voice [25]. Given a speech recognition system that is 100% accurate, the maximum intelligibility of synthesized speech can be obtained. Since enough data is not available to train a speaker-dependent TTS for every dysarthric speaker, an HMM-based adaptive TTS synthesizer is developed. Additionally, the pronunciation of the dysarthric speaker can be corrected.

A flowchart illustrating the phases in an HMM-based adaptive TTS system, namely, training, adaptation and synthesis, is shown in Figure 2.3.[2] Training and adaptation data consists of speech waveforms and corresponding transcriptions. In the training phase, mel-generalized cepstral (MGC) coefficients and pitch ($\log f0$) values are extracted from the speech waveforms. Velocity and acceleration values are also extracted to account for the dynamic nature of speech. Average voice models are trained on features extracted from the training data. To reduce the influence of speaker differences in the training data, speaker adaptive training is performed. In the adaptation phase, average voice models are adapted to the adaptation features. Constrained structural MAP linear regression along with MAP adaptation is used. In the synthesis phase, the test sentence is parsed into its constituent phones. Based on the context, phone HMMs are chosen and concatenated to generate the sentence HMM. MGC coefficients and $f0$ values are generated from the sentence HMM, and the mel log spectrum approximation filter is used to synthesize speech.

Speech data of two normal American male speakers, "bdl" and "rms" from the CMU corpus, is used as the training data for speakers in Nemours database. Speech data of an Indian speaker "ksp" from the CMU corpus is used as the training data for the Indian English dysarthric data. About 1 hour of speech data is available for every speaker in the CMU corpus. Dysarthric speech data is divided into adaptation (80%) and test data (20%). Only sentences in the test data are used in the synthesis phase and the generated synthesized waveforms are used in the subjective evaluation.

2.6 Proposed modifications to dysarthric speech

In the previous two techniques, normal speech data was either used to determine the transformation function or to build an HTS system. The aim of both

---

2 This figure is redrawn from [25]

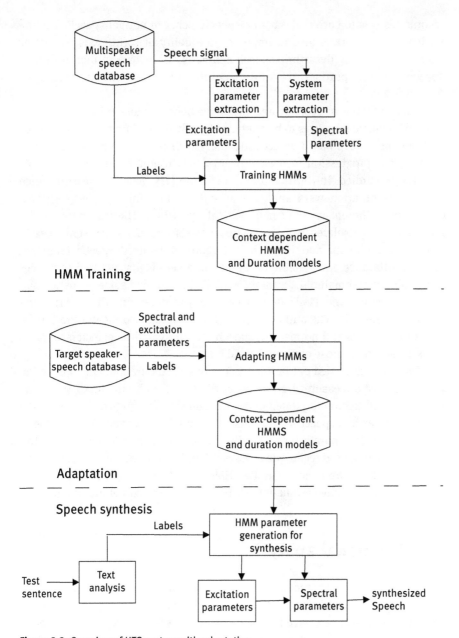

Figure 2.3: Overview of HTS system with adaptation.

techniques was to correct dysarthric speech closer to normal speech. The objective of this work is also to improve the quality of continuous dysarthric speech by bringing dysarthric speech characteristics closer to that of normal speech. For this purpose, dysarthric speech and normal speech have to be first analyzed.

Different types of acoustic analyses have been conducted in the past. The types of acoustic measures to be analyzed largely depend on their application. In Tomik et al. [26], based on acoustic analysis of certain sounds over several months, the progression of dysarthria in Amyotrophic Lateral Sclerosis patients were studied. In Hertrich and Ackermann [27], acoustic measures such as syllable duration were analyzed at syllabic and intrasyllabic segments across four different types of neurological dysarthria. The test materials included a set of isolated words, isolated vowels and nonsense sentences in German. Ackermann and Hertrich [28] reported reduced speech tempo in terms of utterance and syllable durations on a set of 24 structured sentences in German as test materials. Surabhi et al. [29] performed some acoustic analysis on Nemours and TIMIT databases [2, 30] to determine the severity and cause of dysarthria. The authors reported that normalized mean duration for overall phonemes and normalized speech rate of dysarthric speakers were always greater than those of normal speakers.

The analysis carried out in this work is based on duration, which is an important acoustic measure. Durational attributes of both structured (Nemours database) and unstructured sentences (IE dataset) are studied. Based on the analysis, dysarthric speech waveforms are manually corrected. A technique to automatically incorporate these modifications is then developed. The performance of the proposed technique is assessed and compared with that of the formant resynthesis approach and the HMM-based adaptive TTS synthesizer. Results of the subjective evaluation are presented at the end of the chapter.

2.7 Durational analysis

Dysarthria is mostly characterized by slow speech [3, 29]. However, there are studies reporting rapid rate of speech [1, 31]. In Dorze et al. [31], it is observed that dysarthric speakers have higher speaking rates for interrogative sentences. Hypokinetic dysarthria, which is often associated with Parkinson's disease, is characterized by bursts of rapid speech [32]. Hence, it is necessary to assess the speech of people with dysarthria in terms of durational attributes, for the task of speech enhancement.

The corresponding normal speech for the utterances spoken by every dysarthric speaker is available in the Nemours database. The speech of Malayalam speaker "IEm" in the Indic TTS database [22] is considered as the reference normal speech for the Indian dysarthric speaker IE. Using phonemic labeling of the speech data, dysarthric and normal speech are compared based on vowel duration, average speech rate and total utterance duration. In all the subsequent figures, BB-SC refers to speakers in the Nemours database, and IE refers to the Indian speaker with dysarthria.

2.7.1 Vowel durations

The average vowel duration is the average duration of all the vowels pooled together. It is observed that the average vowel duration of dysarthric speech in the databases is longer than their normal speech counterparts [29]. This is illustrated in Figure 2.4. It is observed that the average vowel duration for normal speakers is about 110–150 ms, while it is about 140–320 ms for dysarthric speakers. Longer average vowel durations are noticed especially for speakers BK and RL.

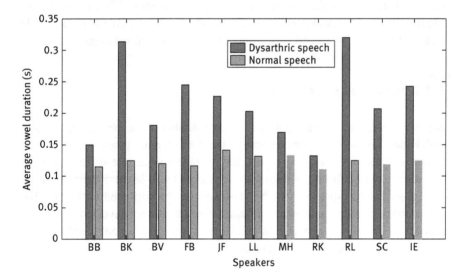

Figure 2.4: Average vowel durations across dysarthric and normal speakers.

Average durations of individual vowels are also analyzed. The number of vowels in the Nemours database and IE dataset are 12 and 16, respectively. It is observed that the average duration for most vowels is longer for dysarthric speech compared to normal speech in most cases. For dysarthric speakers BV and RK, the average duration of vowel "er" is shorter as it is hardly articulated. Surprisingly, for vowel "eh," the average duration is longer for normal speech compared to dysarthric speech for most speakers. We conclude that this may be a characteristic of the normal speaker.

Higher standard deviations are also observed for vowel durations of dysarthric speakers. A larger standard deviation indicates that either the vowel is hardly uttered or is sustained for a longer duration. In Figure 2.5, the estimated probability density function (PDF) of vowel duration for different speakers is plotted. The probability of vowel duration being in a specific range is given by the area of the pdf within that range. It is observed from Figures 2.5(a) and (b) that the degree to which the dysarthric persons' vowel duration differs from their normal counterpart is speaker dependent. The average and standard deviation of vowel durations are closer to normal speech for speaker BB than those for speaker RL.

For speech data of the Indian dysarthric speaker IE, speech data of different normal speakers is considered for comparison. Speech data of an American speaker "rms" from CMU corpus [19] and four different nativities of Indian English (Hindi, Malayalam, Tamil and Telugu) in the Indic TTS corpus [22] are used. From Figure 2.5(c) it is observed that the duration plot of speaker IE is clearly shifted and stretched with respect to that of normal speakers.

2.7.2 Average speech rate

Average speech rate is defined as the average number of phones uttered per second. Figure 2.6 shows the average speech rate for different dysarthric and normal speakers. While for normal speakers, the average speech rate is about 7.5–9.5 phones per second, it ranges from 4 to 8.5 phones for different dysarthric speakers. Lower speech rates for dysarthric speakers can be attributed to the fact that the coordination and movement amongst the articulators involved in speech production are not as smooth and effortless as those for normal speakers. This results in the sustenance of the same sound for a longer time.

2.7.3 Total utterance duration

Total utterance duration is the duration taken to speak/read the same set of sentences by dysarthric and corresponding normal speakers. From Figure 2.7 it is seen that the total utterance duration is longer for dysarthric speakers. The

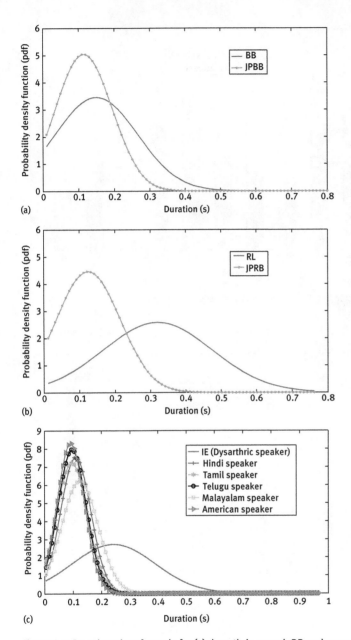

Figure 2.5: Duration plot of vowels for (a) dysarthric speech BB and normal speech JPBB, (b) dysarthric speech RL and normal speech JPRL and (c) Indian dysarthric speech IE and normal speech of other speakers.

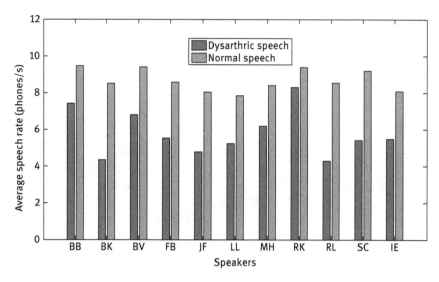

Figure 2.6: Average speech rates across dysarthric and normal speakers.

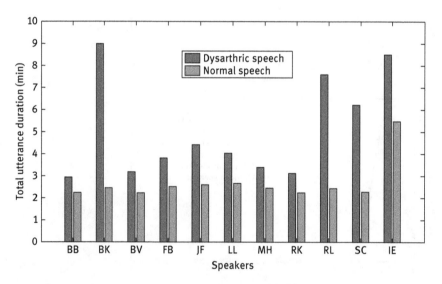

Figure 2.7: Total utterance durations across dysarthric and normal speakers.

normal speaker in the Nemours database takes about 2–3 min to speak multiple sets of 74 six-worded sentences. Most dysarthric speakers take about 2.5–5 min to speak the same set of sentences. Longest utterance durations are evident for speakers BK, RL and SC. This indicates the insertion of phones, intra-utterance pauses and so on while speaking.

Based on the above analysis, the goal is to improve the quality of dysarthric speech by reducing its duration closer to that of normal speech. In Vijayalakshmi and Reddy [33], it is observed that the intelligibility of dysarthric speech in terms of FDA score [21] comes down as phone durations increase. Motivated by this observation, in this work, this durational correction is achieved by modifying dysarthric speech both manually and automatically.

2.8 Manual modifications

For the datasets used in the experiments, it is observed that dysarthric speech has longer average vowel duration, lower speech rate and longer utterance duration compared to normal speech. Increasing the speech rate of the entire dysarthric utterance is not useful; specific corrections need to be made. Dysarthric and normal speech are compared in terms of individual phone segments. The following manual modifications are made to dysarthric speech:

- removing artifacts in the utterance;
- deleting long intra-utterance pauses;
- splicing out steady regions of elongated vowels;
- removing repetitions of phones.

Modifications are made such that the intelligibility of speech is not degraded. Regions of the dysarthric utterance are carefully deleted such that there are no sudden changes in the spectral content. To ensure this, the spectrogram of the waveform is used for visual representation.

Figure 2.8 shows the word-level labeled waveform of an utterance – *The badge is waking the bad*, spoken by dysarthric speaker SC and its manually modified waveform. The first thing to be observed is the reduction in utterance duration in terms of the number of samples. Long pauses, labeled as "pau," are removed from the waveform. It is also observed that the second pause contains some artifacts, which are spliced out in the manual process. The elongation of the vowel "ae" in the word *badge* is also addressed.

For each dysarthric speaker, a set of 12 dysarthric speech waveforms is manually modified. Informal evaluation of original and corresponding manually modified waveforms indicate an improvement in quality in the latter case. Encouraged by this, an automatic technique is developed to replace the manual procedure.

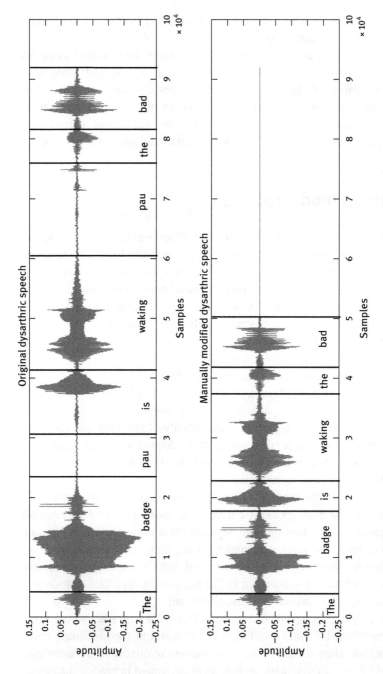

Figure 2.8: Original (top panel) and manually modified (bottom panel) dysarthric speech for the utterance *The badge is waking the bad* spoken by speaker SC.

2.9 Proposed automatic method

An automatic technique is developed to improve the quality of continuous dysarthric speech. A dysarthric speech utterance is compared to the same utterance spoken by a normal speaker at the frame level, and dissimilar regions in the former are spliced out based on certain criteria. DTW algorithm is used to compare the two utterances. The features used to represent an utterance, the DTW algorithm and criteria for deletion, are discussed in the following subsections.

2.9.1 Features used

To represent speech or the content of the spoken utterance, MFCC are used. MFCCs capture the speech spectrum by considering the speech perception nature of humans. Thirteen static MFCCs are extracted from the audio waveform. To capture the dynamic nature of speech, 13 first-order derivatives (velocity) and 13 second-order derivatives (acceleration) of the MFCCs are also determined. Hence, each frame of the speech signal is represented by a 39-dimensional feature vector. Overlapping sliding Hamming windows of 25 ms length and shift of 10 ms are the frame attributes. Cepstral mean subtraction is then performed to compensate for speaker variation [34]. The mean of cepstral coefficients for an utterance is subtracted from each frame cepstral coefficient of that utterance. The mean subtracted MFCCs of the dysarthric and normal speech are the features used for comparison.

2.9.2 Dynamic time warping (DTW)

The DTW is a dynamic programming algorithm that finds the optimal alignment between two temporal sequences [35]. It measures the similarity between two utterance sequences that may vary in timing and pronunciation.

Let the two sequences be:

$$\mathbf{X} = [\vec{x}_1, \vec{x}_2, ..., \vec{x}_i, ..., \vec{x}_m],$$
$$\mathbf{Y} = [\vec{y}_1, \vec{y}_2, ..., \vec{y}_j, ..., \vec{y}_n].$$

(2.1)

$\vec{x}_i$ and $\vec{y}_j$ are the frame vectors of each utterance, respectively. Distance between elements $\vec{x}_i$ and $\vec{y}_j$ is defined by the Euclidean distance,

$$\text{dist}(\vec{x}_i, \vec{y}_j) = \|\vec{x}_i - \vec{y}_j\|$$
$$= \sqrt{(x_{i1} - y_{j1})^2 + (x_{i2} - y_{j2})^2 + \ldots + (x_{id} - y_{jd})^2}, \tag{2.2}$$

where d is the dimensionality of feature vectors, which is 39 in our case. The two sequences can be arranged in the form of an $n \times m$ matrix (or grid) as shown in Figure 2.9. Each element $(\vec{x}_i, \vec{y}_j)$ in the matrix corresponds to the cumulative distance between elements $\vec{x}_i$ and $\vec{y}_j$. The goal of the DTW is to find an optimal path $[(a_1, b_1), (a_2, b_2), \ldots, (a_q, b_q)]$ that minimizes the cumulative distance between the two sequences:

$$\text{DTW}(X, Y) = \min_{p} \left[\sum_{k=1}^{N_p} \text{dist}(\vec{x}_{a_k}, \vec{y}_{b_k}) \right], \tag{2.3}$$

where p is the number of possible paths, and N_p is the number of cells in the pth path.

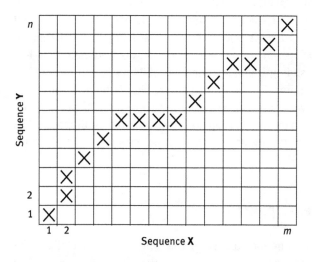

Figure 2.9: DTW grid.

To restrict the space of the warping paths, the following constraints are included [35]:
- Monotonicity: The indices of the path should be monotonically ordered with respect to time: $a_{k-1} \leq a_k$ and $b_{k-1} \leq b_k$.
- Continuity: Only one step movement of the path is allowed: $a_{k-1} - a_k \leq 1$ and $b_{k-1} - b_k \leq 1$.
- Boundary conditions: The first and last vectors of X and Y are aligned to each other: $a_1 = b_1 = 1$ and $a_q = m, b_q = n$.

An example of an optimal path is shown in Figure 2.9. Diagonal regions indicate similarity between the sequences, horizontal regions indicate elongation in sequence X and vertical regions indicate elongation in sequence Y. The dysarthric utterance is the test sequence (X) and the normal utterance is the reference sequence (Y). To obtain the optimal path between the feature vectors of dysarthric and normal speech utterances, the DTW algorithm is used. Then the slope of the path is calculated. Wherever the slope is zero, the corresponding frames are deleted from the waveform. However, discontinuities are perceived in the resultant waveform. Therefore, certain criteria are considered before deletion of frames.

2.9.3 Frame threshold

An initial criterion for deletion of speech frames is the allowance of some amount of elongation in the dysarthric utterance. Whenever the horizontal path is steady or the slope of the DTW path is zero for a minimum number of frames (*frameThres*), corresponding speech frames are spliced out.

2.9.4 Short-term energy criterion

At concatenation points, that is, frames before and after deleted regions, it is observed that artifacts are introduced if the energy difference between frames is high. Hence, when deleting frames, it is necessary to ensure that there is no sudden change in energy at the concatenation points. An additional criterion for deletion is the short-term energy (STE – square of L2-norm of speech segment) difference. Frames are deleted whenever STE difference is less than a certain limit, referred to as *STEThres*. Algorithm 2.1 is proposed in this work to select speech frames for deletion.

The automatic modification method is referred to as the DTW + STE modification technique. This automatic technique is illustrated in Figure 2.10. Thresholds *frameThres* and *STEThres* are varied from 4 to 10 and 0.3 to 2.5, respectively. It is empirically determined that *frameThres* and *STEThres* values of 6 and 0.5, respectively, seem to work the best. Systems using these best thresholds are used for subjective evaluation. Experiments are also performed using normalized STE difference as a threshold. The performance of using normalized STE difference as a threshold is on par with that using *STEThres*.

Figure 2.11 shows the DTW paths of a sample utterance of dysarthric speaker BK before and after automatic modifications compared with respect to the same utterance of normal speaker JPBK. Comparing Figures 2.11(a) and (b),

Algorithm 2.1: Determining the frames for deletion.

Input: Frame numbers corresponding to a deletion region (*delFrames*) after thresholding with *frameThres*

Let $delFrames = [g_1, g_2, ..., g_k]$

Steps:

1. $consideredFrames = [g_1, g_2, ..., g_k, g_{k+1}]$

2. Calculate STE for all frames in *consideredFrames*

3. Calculate absolute STE difference for every pair of frame combinations

4. Identify frame pairs with $STE \leq STEThres$

5. From the identified frame pairs, choose the frame pair for which maximum frames can be deleted. If the chosen frame pair is (g_i, g_j) with $i < j$, then the frames to be deleted are $g_{i+1}, ..., g_{j-1}$.

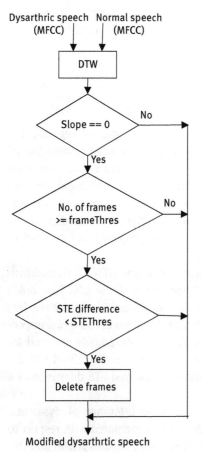

Figure 2.10: Flowchart of automatic (DTW + STE) method to modify dysarthric speech.

it can be concluded that the modified speech is closer to that of normal speech as indicated by the diagonal DTW path in Figure 2.11(b). It also results in a significant reduction in the number of frames or duration of the utterance.

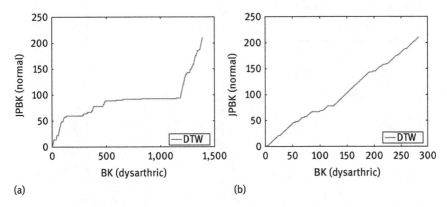

(a)   (b)

Figure 2.11: DTW paths of an utterance of speaker BK between (a) original dysarthric speech and normal speech and (b) modified dysarthric speech and normal speech.

An example of the same utterance spoken by dysarthric speaker BK and that modified using the DTW + STE method is given in Figure 2.12. It is observed that pauses and extra (inserted) phones (PI) present in the original dysarthric speech are removed in the automatic technique. Overall there is a reduction in the number of samples or duration of the dysarthric utterance. The total utterance duration of the original and automatically modified dysarthric speech are given in Figure 2.13. The total utterance duration of the modified speech is closer to normal speech, except in the case of speakers BB and RK, where the reduced utterance duration compared to normal speech is an indication of some amount of elongation present in the normal speech. Moreover, for these two speakers, dysarthric speech is closer to normal speech in terms of the duration attributes.

The DTW + STE method not only results in a durational reduction, but it also improves the intelligibility of dysarthric speech by removing insertions, long pauses and elongations. Subjective evaluation is performed to compare the quality of the automatically modified and original dysarthric speech.

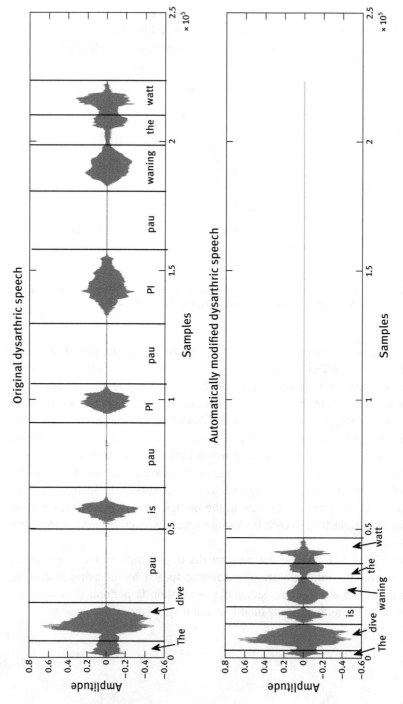

Figure 2.12: Original (top panel) and automatically modified (bottom panel) dysarthric speech for the utterance *The dive is waning the watt* spoken by speaker BK. PI – phone insertions.

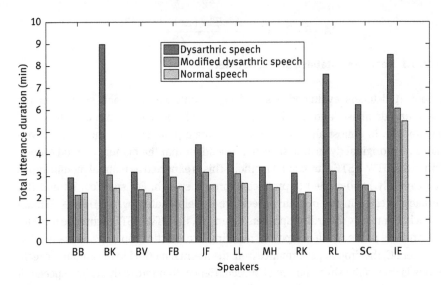

Figure 2.13: Total utterance durations across dysarthric speech, modified dysarthric speech and normal speech.

2.10 Performance evaluation

To evaluate the techniques implemented for enhancement of dysarthric speech, sub-jective evaluation is carried out. To assess the proposed modification techniques, a pairwise comparison (PC) test is performed. To compare intelligibility across different methods, a word error rate (WER) test is conducted. Naive listeners are used in sub-jective tests rather than expert listeners. This is done to assess how a naive listener, who has little or no interaction with dysarthric speakers, evaluates the quality of dys-arthric speech. All the listening tests are conducted in a noise-free environment.

2.10.1 Pairwise comparison tests

To compare the quality of speech modified by the proposed techniques and the orig-inal dysarthric speech, a PC test is conducted [36]. The PC test consists of two subt-ests: "A–B" and "B–A." In the "A–B" test, speech output of system "A" is played first and then that of system "B," and vice versa in the "B–A" test to remove any bias in listening. In both the tests, "A" is the modified speech and "B" is the original speech. Preference is always calculated in terms of the audio sample played first. The overall preference for system "A" against system "B" is given by a combined score "A–B + B–A" and is calculated using the following formula:

$$"A - B + B - A" = \frac{"A - B" + (100 - "B - A")}{2}$$

2.10.1.1 Nemours database and IE dataset

About 11 listeners evaluated a set of eight sentences for each speaker in the Nemours database and IE dataset. Results of the evaluation are shown in Figure 2.14. In almost all cases, results indicate a preference for the modified versions over original dysarthric speech. It is evident that the manual method outperforms the DTW + STE (automatic) method. This is expected as manual modifications are carefully hand-crafted to produce improved speech. For speakers BK and BV, artifacts present in the original speech are not eliminated by the DTW + STE approach. For mildly dysarthric speakers BB and IE, the DTW + STE technique introduces artifacts in the modified speech, resulting in a drastic drop in performance. For speaker SC, the drop in performance from the manual method to the automatic technique is due to the slurry nature of speech. Hence, to improve dysarthric speech, it is essential to identify the specifics of dysarthria for individual speakers. Nonetheless, the performance of both approaches is almost on par for speakers JF, LL, FB, MH, RL and RK who are mild-to-severely dysarthric.

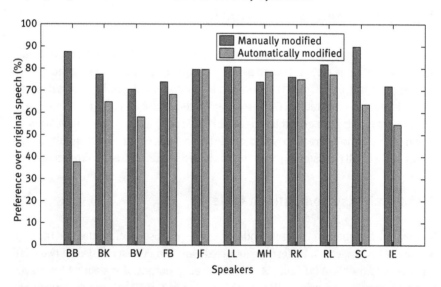

Figure 2.14: Preference for manually and automatically (DTW + STE)-modified speech over original dysarthric speech of different speakers (Nemours database and IE dataset).

Modified dysarthric speech using the formant resynthesis technique was also evaluated. They were compared with the original and automatically modified

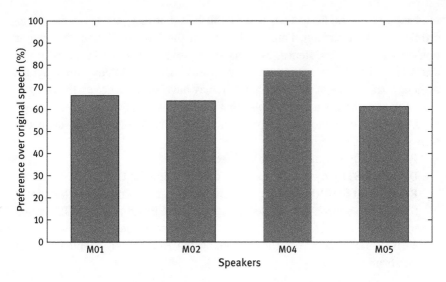

Figure 2.15: Preference for automatically (DTW + STE)-modified speech over original dysarthric speech of different speakers (TORGO database).

(DTW + STE) dysarthric speech in two separate PC tests. About 10 listeners evaluated a set of eight sentences for each speaker in each test. The formant resynthesis technique was clearly not preferred; the competing technique scored considerably better at 82% in both the tests.

2.10.1.2 TORGO database

About 10 listeners evaluated a set of eight sentences for each speaker considered in the TORGO database. Results of the PC test are shown in Figure 2.15. It is seen that there is a clear preference for speech modified using the DTW + STE technique over the original dysarthric speech. The preference is lowest for speaker M05, who is moderate-severely dysarthric; the other speakers are severely dysarthric. Highest preference is for speaker M04. Speaker M04 stutters while speaking, and most repetitions are observed to be eliminated using the DTW + STE technique.

2.10.2 Intelligibility test

A WER test was conducted to evaluate intelligibility across different systems. Nemours database and IE dataset were alone considered in this test. Based on the feedback on the PC tests, it was established that recognizing words in dysarthric speech is difficult. Further, the text in the Nemours database contains nonsensical sentences. Hence, listeners were provided the text in the WER test and were asked to enter the number of words that were totally unintelligible. Although knowledge of the pronunciation may have an influence on the recognition of the text, this is a uniform bias that is present when evaluating all the systems. About 10 evaluators participated in the test. The following types of speech were used in the listening tests:

P (original): original dysarthric speech

Q (DTW + STE): dysarthric speech modified using the DTW + STE method

R (Formant Synth): output speech of the formant resynthesis technique

S (HTS-in): speech synthesized using the HMM-based adapted TTS for text in the database not used for training (held-out sentences)

T (HTS-out): speech synthesized using the HMM-based adapted TTS for text from the web

Results of the WER test are presented in Figure 2.16. It is seen that the intelligibility of formant resynthesis technique is poor for all the speakers. For a majority of speakers, the DTW + STE method has a higher WER compared to original dysarthric speech. In almost all the cases, WER of HMM-based adaptive synthesizer on held-out sentences (sentences not used during training) is comparatively high with respect to the original dysarthric speech. For speakers in Nemours database, the intelligibility of sentences synthesized from the web is quite poor compared to that of held-out sentences. This trend is the reverse for the Indian dysarthric speaker IE. This is because the structure of the held-out

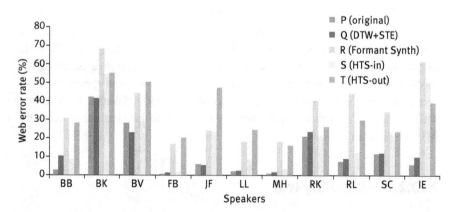

Figure 2.16: Word error rates of different types of speech across dysarthric speakers (Nemours database and IE dataset).

sentences in Nemours database is similar to the one used during training. This is not the case in the Indian dysarthric dataset, in which the training and held-out sentences are unstructured. Overall, the intelligibility of original dysarthric speech does not increase. However, the DTW + STE modified speech has the lowest WER for speakers BK, BV and JF. For speaker RK, the intelligibility of HMM-based adaptive synthesized speech is on par with that of original dysarthric speech. It is observed that some mispronunciations in dysarthric speech get corrected for sentences synthesized using the HMM-based adaptive synthesizer. This suggests that the approach used to increase intelligibility largely depends on the type and severity of dysarthria.

2.11 Analogy to speech synthesis techniques

The HMM-based adaptive synthesizer [25] is a statistical parametric speech synthesizer. The synthesized speech output of the HMM-based synthesizer lacks the voice quality of the dysarthric speaker.

The DTW + STE technique can be considered analogous to a unit selection speech (USS) synthesizer [37]. In USS, subword units are concatenated together to produce speech. Units are selected based on certain target and concatenation costs. In a similar manner, in the DTW + STE method, frames to be concatenated are selected based on STE difference criteria. Although the speech output of the DTW + STE method has discontinuities, it sounds more natural and preserves the voice characteristics of the dysarthric speaker.

2.12 Summary

Dysarthric speech is enhanced to assist people with dysarthria communicate effectively. In addition to using the Nemours and TORGO databases for experiments, continuous and unstructured speech of a native Indian having dysarthria is collected. Normal speech characteristics are transplanted onto dysarthric speech to improve the latter's quality while preserving the speaker's voice characteristics. A durational analysis is first performed across dysarthric and normal speech. Dysarthric speech waveforms are manually modified based on the analysis. The DTW + STE technique is developed to automatically correct durational attributes of dysarthric speech closer to that of normal speech. This technique is compared with two other techniques available in the literature, namely, a formant resynthesis technique and a HMM-based adaptive TTS technique. The intelligibility of dysarthric speech modified using different techniques is studied. Although the DTW + STE technique does not increase intelligibility for most speakers, there is a clear preference for the modified dysarthric speech using the proposed technique, that is, there is an improvement in the perceptual quality of the modified dysarthric speech. This highlights the importance of duration in perceptual speech quality. This kind of modification may be used as a preprocessing step to enhance dysarthric speech.

Although the DTW + STE technique does not require labels or aligned boundaries, it makes use of the normal speech utterance as reference. In this work, only insertion of sounds is considered; deletion and substitution of phones are not addressed. Besides the analysis of durational attributes, other attributes affecting the speech of a person with dysarthria can be explored. To build an efficient and practical device for enhancing dysarthric speech, a person-centric approach needs to be adopted.

References

[1]   American Speech Language Hearing Association, "Dysarthria," www.asha.org/public/speech/disorders/dysarthria/, [last accessed 22-2-2017].

[2]   Menendez-Pidal, X., Polikoff, J., Peters, S., Leonzio, J., and Bunnell, H. "The Nemours database of dysarthric speech," in Fourth International Conference on Spoken Language (ICSLP), Philadelphia, USA, October 1996, pp. 1962–1965.

[3]   Rudzicz, F., Namasivayam, A. K., and Wolff, T. The TORGO database of acoustic and articulatory speech from speakers with dysarthria. Language Resources and Evaluation, 46(4), 523–541, 2012.

[4]   Prakash, A., Reddy, M. R., and Murthy, H. A. Improvement of Continuous Dysarthric Speech Quality, Speech and Language Processing for Assistive Technology (SLPAT), San Francisco, USA, September, 2016, 43–49.

[5]   Deller, J., Hsu, D., and Ferrier, L. On the use of hidden Markov modelling for recognition of dysarthric speech. Computer Methods and Programs in Biomedicine, 35(2), 125–139, 1991.

[6]   Jayaram, G., and Abdelhamied, K. Experiments in dysarthric speech recognition using artificial neural networks. Rehabilitation Research and Development, 32(2), 162–169, May 1995.

[7]   Hosom, J. P., Kain, A. B., Mishra, T., van Santen, J. P. H., Fried-Oken, M., and Staehely, J. "Intelligibility of modifications to dysarthric speech," in International Conference on Acoustics, Speech, and Signal Processing (ICASSP), Hong Kong, China, April 2003, pp. 924–927.

[8]   Rudzicz, F. Acoustic transformations to improve the intelligibility of dysarthric speech, 2nd Workshop on Speech and Language Processing for Assistive Technologies (SLPAT), Edinburgh, UK, July 2011, 11–21.

[9] Kain, A. B., Hosom, J.-P., Niu, X., van Santen, J. P., Fried-Oken, M., and Staehely, J. Improving the intelligibility of dysarthric speech. Speech Communication, 49(9), 743–759, 2007.

[10] Rudzicz, F. Adjusting dysarthric speech signals to be more intelligible. Computer Speech & Language, 27(6), 1163–1177, 2013.

[11] Yakcoub, M. S., Selouani, S. A., and O'Shaughnessy, D. "Speech assistive technology to improve the interaction of dysarthric speakers with machines," in 3rd International Symposium on Communications, Control and Signal Processing (ISCCSP), Malta, March 2008, pp. 1150–1154.

[12] Dhanalakshmi, M., and Vijayalakshmi, P. "Intelligibility modification of dysarthric speech using HMM-based adaptive synthesis system," in 2nd International Conference on Biomedical Engineering (ICoBE), Penang, Malaysia, March 2015, pp. 1–5.

[13] Saranya, M., Vijayalakshmi, P., and Thangavelu, N. "Improving the intelligibility of dysarthric speech by modifying system parameters, retaining speaker's identity," in International Conference on Recent Trends In Information Technology (ICRTIT), Chennai, India, April 2012, pp. 60–65.

[14] Kain, A., Niu, X., Hosom, J., Miao, Q., and van Santen, J. P. H. Formant re-synthesis of dysarthric speech, Fifth ISCA ITRW on Speech Synthesis, Pittsburgh, USA, June 2004, 25–30.

[15] Kim, H., Hasegawa-Johnson, M., Perlman, A., Gunderson, J., Huang, T. S., Watkin, K., and Frame, S. "Dysarthric speech database for universal access research." in Annual Conference of the International Speech Communication Association, INTERSPEECH, Brisbane, Australia, September 2008, pp. 1741–1744.

[16] M. C. T. A, N. T, and V. P. "Dysarthric speech corpus in Tamil for rehabilitation research," in IEEE Region 10 Conference (TENCON), Singapore, November 2016, pp. 2610–2613.

[17] Wikipedia, "Arpabet," https://en.wikipedia.org/wiki/Arpabet, [last accessed 22-2-2017].

[18] Aswin Shanmugam, S., and Murthy, H. A. "Group delay based phone segmentation for HTS," in National Conference on Communication (NCC), Kanpur, India, February 2014, pp.1–6.

[19] Kominek, J., and Black, A. W. The CMU arctic speech databases, 5th ISCA Speech Synthesis Workshop, June 2004, 223–224.

[20] Carnegie Mellon University, "The CMU pronunciation dictionary," www.speech.cs.cmu.edu/cgi-bin/cmudict, [last accessed 22-2-2017].

[21] Enderby, P. Frenchay Dysarthria Assessment. International Journal of Language & Communication Disorders, 15(3), 165–173, December 2010.

[22] Baby, A., Thomas, A. L., N.L., Nishanthi and Murthy, H. A. "Resources for Indian languages," in Community-based Building of Language Resources (International Conference on Text, Speech and Dialogue), Brno, Czech Republic, September 2016, pp. 37–43.

[23] Reynolds, D. A., Quatieri, T. F., and Dunn, R. B. Speaker verification using adapted gaussian mixture models. Digital Signal Processing, 10(1–3), 19–41, January 2000.

[24] Rabiner, L., and Schafer, R. Theory and Applications of Digital Speech Processing, 1st, 1em plus 0.5em minus 0.4em Upper Saddle River, NJ, USA, Prentice Hall Press, 2010.

[25] Yamagishi, J., Nose, T., Zen, H., Ling, Z. H., Toda, T., Tokuda, K., King, S., and Renals, S. Robust speaker-adaptive HMM-based text-to-speech synthesis. IEEE Transactions on Audio, Speech, and Language Processing, 17(6), 1208–1230, August 2009.

[26] Tomik, B., Krupinski, J., Glodzik-Sobanska, L., Bala-Slodowska, M., Wszolek, W., Kusiak, M., and Lechwacka, A. Acoustic analysis of dysarthria profile in ALS patients. Journal of the Neurological Sciences, 169(2), 35–42, 1999.

[27] Hertrich, I., and Ackermann, H. Acoustic analysis of durational speech parameters in neurological dysarthrias, From the Brain to the Mouth, ser. Neuropsychology and Cognition. 1em plus 0.5em minus 0.4em Springer Netherlands, 1997, Vol. 12, 11–47.

[28] Ackermann, H., and Hertrich, I. Speech rate and rhythm in cerebellar dysarthria: An acoustic analysis of syllabic timing. Folia Phoniatrica et Logopaedica, 46(2), 70–78, 1994.

[29] Surabhi, V., Vijayalakshmi, P., Lily, T. S., and Jayanthan, R. V. Assessment of laryngeal dysfunctions of dysarthric speakers, IEEE Engineering in Medicine and Biology Society, Minnesota, USA, September 2009, 2908–2911.

[30] Garofolo, J. S., Lamel, L. F., Fisher, W. M., Fiscus, J. G., Pallett, D. S., and Dahlgren, N. L., "DARPA TIMIT acoustic phonetic continuous speech corpus CDROM," 1993. [Online]. Available: http://www.ldc.upenn.edu/Catalog/LDC93S1.html

[31] Dorze, G. L., Ouellet, L., and Ryalls, J. Intonation and speech rate in dysarthric speech. Journal of Communication Disorders, 27(1), 1–18, 1994.

[32] Johnson, A., and Adams, S. Nonpharmacological management of hypokinetic dysarthria in Parkinson's disease. Journal of Geriatrics and Aging, 9(1), 40–43, 2006.

[33] Vijayalakshmi, P., and Reddy, M. "Assessment of dysarthric speech and analysis on velopharyngeal incompetence," in 28th Annual International Conference of the IEEE Engineering in Medicine and Biology Society (EMBS), New York, USA, August 2006, pp. 3759–3762.

[34] Westphal, M. "The use of cepstral means in conversational speech recognition," in European Conference on Speech Communication and Technology (EUROSPEECH), Rhodes, Greece, September 1997, pp. 1143–1146.

[35] Sakoe, H., and Chiba, S. Dynamic programming algorithm optimization for spoken word recognition. IEEE Transactions on Acoustics, Speech, and Signal Processing, 26(1), 43–49, 1978.

[36] Salza, P., Foti, E., Nebbia, L., and Oreglia, M. MOS and pair comparison combined methods for quality evaluation of text to speech systems. in Acta Acustica, 82, 650–656, 1996.

[37] Hunt, A. J., and Black, A. W. "Unit selection in a concatenative speech synthesis system using a large speech database," in International Conference on Acoustics and Speech Signal Processing (ICASSP), Atlanta, USA, May 1996, pp. 373–376.

P. Vijayalakshmi, M. Dhanalakshmi and T. Nagarajan

3 Assessment and intelligibility modification for dysarthric speech

Abstract: Dysarthria is a motor speech disorder that is often associated with irregular phonation and amplitude, incoordination and restricted movement of articulators. This condition is caused by cerebral palsy, degenerative neurological disease and so on. The pattern of speech impairment can be determined by the amount of compromise detected in the muscle groups. That is, the dysarthrias have global effect rather than focal effects on speech production systems of phonation, articulation and resonance.

Clinically, assessment of dysarthria is carried out using perceptual judgment by experienced listeners. One of the limitations of perceptual assessment is that it can be difficult even for highly trained listeners to differentiate the multiple dimensions of dysarthric speech, as dysarthria has multisystem dysregulation. Although many researchers are involved in developing assistive devices, acoustic analyses are carried out on each of the subsystems independent of each other. As dysarthria affects the speech system globally, a multidimensional approach is required for the assessment and an associated intelligibility improvement system to develop an assistive device.

This chapter will describe the significance and methods to develop a detection and assessment system by analyzing the problems related to laryngeal, velopharyngeal and articulatory subsystems for dysarthric speakers, using a speech recognition system and relevant signal-processing-based techniques. The observations from the assessment system are used to correct and resynthesize the dysarthric speech, conserving the speaker's identity, thereby improving the intelligibility. The complete system can detect the multisystem dysregulation in dysarthria, correct the text and resynthesize the speech, thus improving the lifestyle of the dysarthric speaker by giving them the freedom to communicate easily with the society without any human assistance.

Keywords: dysarthria, TTS, velopharyngeal incompetence, intelligibility improvement, communication aid

https://doi.org/10.1515/9781501501265-004

3.1 Introduction

Dysarthria as the name suggests is a speech disorder involving abnormal (dys) articulation (arthr) of sounds. Dysarthrias are complex disorders as they represent a variety of neurological disturbances that can potentially affect a single or every component of speech production. The possible neurological disturbances include Parkinson's disease, multiple sclerosis, degenerative neurological diseases, stroke or brain injury [1–4] and so on. Depending on the location and extent to which the nervous system is affected, the severity of the disorder differs. Dysarthrias affect a single subsystem of speech or the disruption may be distributed over components of respiratory, velopharyngeal and articulatory subsystems. The degree to which the muscle groups are compromised due to dysarthria determines the particular pattern of speech impairment.

Dysarthrias are often associated with irregular phonation and amplitude, incoordination and restricted movement of articulators and affect multiple dimensions of spoken language such as speech quality, intelligibility and prosody impeding effective and efficient communication [1–3]. Dysarthrias are classified as (i) spastic, (ii) hyperkinetic, (iii) hypokinetic, (iv) flaccid, (v) ataxic and (vi) mixed dysarthria [2, 4], based on the location of the damage in the central or peripheral nervous system. All types of dysarthrias affect articulation of consonants, causing slurring of speech. In very severe cases, vowels are also distorted. Intelligibility varies greatly depending on the damage to the nervous system. Nasalization of voiced sounds is frequently present. Precisely, dysarthrias are multisystem dysregulation of speech and to improve upon the effective communication of dysarthric speakers, assessment and enhancement of speech produced by each of the subsystems of speech is mandatory.

Speech signal generated by a time-varying vocal tract system excited by time-varying excitations is either temporally or spectrally a reflection or representation of the functioning of individual components of speech production. Therefore any acoustic deviations of speech signal will depict the malfunctioning of the corresponding components (speech source and subsystems) and can be interpreted as deviations from their normal counterparts. Signal processing techniques (temporal as well as spectral) assist us in capturing these acoustic deviations in source and system characteristics, and assessment of disordered speech and the subsequent correction at the signal level is made easier. Along with signal-processing techniques various speech technologies such as speech recognition, speech synthesis and speech enhancement techniques support and enhance the speaking capability of persons with speech disorders.

3.2 Effect of dysarthria on speech

Depending on the subsystems of speech that is affected, dysarthric speech may have the following characteristics: (i) pitch breaks, voice tremor, monoloudness or excess loudness variation and monopitch or monotone when the laryngeal subsystem dysfunctions; (ii) hypernasality due to velopharyngeal incompetence and (iii) irregular articulatory breakdown, repetition of phonemes, distortion of vowels due to defective articulatory system. At the signal level, dysarthric speakHerturbations like wide variations in pitch jitter and amplitude shimmer, intensity variations and so on. Similarly, the problems of velopharyngeal functions reflect in the form of addition of resonances or antiresonances, widening of formant bandwidth and so on. The articulatory dysfunctions get reflected in the form of variations in speaking rate, articulatory configurations of vowels and consonants, resulting in unintelligible speech. As dysarthria has concurrent disorders of respiration, phonation, articulation and resonance, dysarthric speech may have a combination of one or more of the above characteristics.

The field of assessment and intelligibility modification of dysarthric speech has seen many approaches from perceptual judgement and analysis based on instrumentation, to acoustic analysis, to synthesis of disordered speech with improved intelligibility. This chapter is restricted to description of various acoustic analysis that can be carried out for the assessment of dysarthric speech and the successful approaches for the intelligibility modification of dysarthric speech.

Acoustic analysis can be informative since it provides quantitative analyses that carry potential for subsystem description and holds key details regarding speech rate, articulatory configuration for vowels and consonants, rates of change in overall configuration of vocal tract and so on. It helps determining the correlates of perceptual judgements of intelligibility, quality and type of dysarthria. Therefore, acoustic analyses can be a valuable complement to perceptual judgement. Although perceptual judgement has been the primary means for classification and description, the reliability and validity are questionable as these are performed by different specialists with no common training in perceptual rating [5]. Also, perceptual judgements alone cannot discriminate between disruptions that occur simultaneously in two or more speech subsystems.

However, the progress in acoustic studies of dysarthria has been slow owing to several factors, including (1) the relatively modest research effort given to neurogenic speech disorders and (2) the difficulty of acoustic analysis for speakers who may have phonatory disruptions, hypernasality, imprecise articulation and other properties that confound acoustic description [5]. Influence of one subsystem on the other may disrupt the analysis and will not be able to come to a conclusion on effects. For instance, introduction of additional formants may be due to velopharyngeal or articulatory dysfunction. Instead of analyzing the combined effects due to all the subsystems, it is better to divide and conquer the assessment process. That is, analyze the subsystems individually, understand the effects of each of these systems separately and combine the results for the assessment of the multisystem dysregulation.

3.3 Divide and conquer approach

To divide and conquer the assessment of various subsystems involved in the production of speech, let us look at one subsystem at a time exclusively independent of the other and analyze the features that are unique to the respective subsystem. In order to do so, the speech system is divided into three major subsystems, namely: (i) velopharyngeal, (ii) laryngeal and (iii) articulatory. The following sections describe the effect of dysfunctions in each of these subsystems separately and the acoustic cues to detect and analyze each one of them and the signal processing and usage of various speech technologies to conquer the assessment and intelligibility modification.

3.3.1 Speech corpora

To describe and to substantiate the effect of various techniques presented in this chapter, in the assessment and intelligibility modification of dysarthric speech, four different speech corpora are used. The description of each of the corpus is given below.

Forty-eight speakers with cleft lip and palate are chosen for the collection of speech data. Among the 48 speakers, 33 speakers are identified as preoperative (unrepaired) cleft-lip and palate speakers, and the remaining 15 speakers are identified as postoperative (with repaired cleft palate) speakers [6]. As vowel nasalization is considered for analysis, the vowels /a/, /i/, /u/ are collected from these preoperative and postoperative as well as from 30 normal speakers for comparison. All speech data are recorded with a head mounted microphone with a frequency response of 20 Hz to 20 kHz, with a sampling rate of 16 kHz.

Assessment of dysarthria is carried out on a database consisting of 10 dysarthric speakers with varying degree of dysarthria and one normal speaker's speech data from the Nemours database of dysarthric speech [7]. The database is designed to test the intelligibility of dysarthric speech before and after enhancement by various signal processing methods. It can also be used to analyze the general characteristics such as the production error patterns of the dysarthric speech. The Nemours corpus is a collection of 740 short nonsense sentences; 74 sentences spoken by each of the dysarthric speaker. The nonsense sentences are of the form "The X is Ying the Z" are syntactically correct and semantically vacuous. Out of the 74 sentences, the first 37 sentences contain the same words as the last 37 sentences with the position of X and Z reversed.

Time-aligned phonetic transcriptions are available for all the dysarthric speakers' speech data in the Nemours corpus. The database also contains Frenchay dysarthria assessment (FDA) [8] scores for nine dysarthric speakers as one speaker is affected by mild dysarthria. FDA is a well-established test for the diagnosis of dysarthria. The test is divided into 11 sections, namely, reflex, palate, lips, jaw, tongue, intelligibility and so on. Each dysarthric speaker is rated on a number of simple tasks. FDA is a 9-point scale, with scores ranging from 0 to 8. In FDA, a score of "8" represents normal function and "0" represents no function.

Apart from this, for comparison normal speech data from TIMIT [9] speech corpus is used. Further, to train HMM-based adaptive speech synthesis system, the CMU Arctic database [10] is used.

The dysarthric speakers considered for this work are affected mostly with spastic dysarthria. The characteristics of spastic dysarthria are [11]: (a) strained or strangled voice; (b) low pitch with pitch breaks occurring in some cases; (c) occurrence of hypernasality typically but not severe; (d) reduced range of movement, tongue strength, speech rate and voice onset time for stops and (e) increase in phoneme-to-phoneme transitions in syllable and word duration and in voicing of unvoiced stops.

3.3.2 Velopharyngeal incompetence

Air passing through the nasal cavity due to inadequate velopharyngeal closure leads to the introduction of inappropriate nasal resonances into the oral sounds, resulting in hypernasal (nasalized) speech. Velopharyngeal dysfunction may result in speech with reduced quality and intelligibility. Some of the effects of velopharyngeal dysfunction are (i) hypernasality, (ii) nasal air emission during consonant production, (iii) weak or omitted consonants and so on. These effects are reflected as acoustic deviation in hypernasal speech in the form of broadening of formant bandwidth, inclusion of nasal formants and antiformants, formant frequency shift, reduced formant amplitude and so on. With the inclusion of addition or deletion of formants, formant frequency shift may be due to irregularities in articulation as well. Therefore, in order to understand the acoustic deviations exclusively due to velopharyngeal dysfunction, speech affected by velopharyngeal dysfunction is analyzed and the results are extended to multisystem dysregulation. A defective velopharyngeal mechanism and the resulting hypernasal speech can be caused by anatomical defects (cleft palate) or hearing impairment (insufficiency) other than due to central or peripheral nervous system damage (incompetence). Therefore, initial analysis is carried out on speakers with unrepaired cleft palate and results are applied to the speech data of dysarthric speakers.

Vowels and voiced oral consonants are in general prone to have hypernasality. Vowel nasalization can be considered over three categories of vowels namely front (/i/), mid (/a/) and back (/u/), since as opposed to consonants these vowels can be sustained for longer time. Initial acoustic analysis on nasalized vowels, vowels modified by introducing various nasal resonances and hypernasal speech, revealed the fact that additional nasal resonances introduced around 250 Hz and 1000 Hz, due to oral-nasal coupling, play a significant role in nasalization [12–13]. This is further verified by perceptual analysis that shows that introduction of nasal resonance in the low-frequency region around 250 Hz that lies below the first formant of the three vowels /a/, /i/, and /u/ is found to be an important and consistent cue for the assessment of velopharyngeal dysfunction (insufficiency as well as incompetence) and vowel nasalization. Further, for the vowels /i/ and /u/, it is noticed that there is a widening of formant bandwidth around the first formant as these vowels have their first oral formant also around 300 Hz. The introduction of nasal formant very close to the oral formant results in the widening of formant bandwidth and is not resolved consistently.

As introduction of nasal formant below first oral formant is the consistent cue for detection of hypernasality, a spectrum estimation technique with a better frequency resolving power is mandatory. The magnitude spectrum derived using Fourier transform cannot be used directly for formant extraction due to the presence of pitch harmonics. Although cepstrum-based technique smoothen the magnitude spectrum, the smaller size of the cepstral lifter, used for smoothening, diminishes the resolving power. Linear prediction (LP)-based analysis technique can, by default, provide a smoothened magnitude spectrum. However, the resolving power and the correctness of LP-based spectral estimation technique are vulnerable to the prediction order. These conventional methods do not resolve the closely spaced frequencies due to (i) poor frequency resolution and (ii) influence of adjacent roots. Modified group delay-based formant extraction technique can alleviate the problem of poor frequency resolution, as group delay function exhibits an additive property unlike magnitude spectrum derived using Fourier transform.

3.3.2.1 The group delay function

The negative derivative of the Fourier transform phase $\theta(\omega)$ is defined as group delay and is given by

$$\tau(\omega) = -\frac{\partial\theta(\omega)}{\partial\omega} \tag{3.1}$$

(In this section, the continuous and discrete frequency variables are used interchangeably, depending upon the requirement.) The group delay function can be derived directly from the signal as described in [14, 15] as given below where $X(K)$ and $Y(K)$ are the N-point DFTs of the sequences $x(n)$ and $n^*x(n)$ respectively:

$$\tau_p(K) = \frac{X_R(K).Y_R(K) + X_I(K).Y_I(K)}{|X(K)|^2}, K = 0, 1, ..., N-1 \tag{3.2}$$

To reduce the spiky nature of group delay function that is due pitch peaks, noise and windowing effects, the group delay function is modified [14] as given below

$$\tau_p(K) = \text{sign} \left| \frac{X_R(K).Y_R(K) + X_I(K).Y_I(K)}{|S(K)|^{2\gamma}} \right|^{\alpha} \tag{3.3}$$

To derive a smoothed group delay spectrum, the value of α and γ should be less than one. For this study, the parameters α and γ are tuned to 0.6 and 0.9, respectively. The group delay spectrum exhibits additive property unlike the magnitude spectrum. If

$$H(\omega) = H_1(\omega).H_2(\omega) \tag{3.4}$$

then the group delay function $\tau_h(\omega)$ is given by

$$\tau_h(\omega) = \frac{\partial(\theta(H(\omega)))}{\partial\omega} \tag{3.5}$$

$$\tau_h(\omega) = \tau_{h1}(\omega) + \tau_{h2}(\omega)$$

Figure 3.1 [6] demonstrates the additive property exhibited by group delay function and its resolving power. Further, the value of the group delay function of a pole at its own angular frequency ω_k is influenced by the rest of the roots and the influence $\tau_k^i(\omega_k)$ is given by

$$\tau_k^i(\omega_k) = \sum_{p=1\&p\neq k}^{P} \tau_p(\omega_k) + \sum_{n=1\&n\neq k}^{N} \tau_n(\omega_k) \tag{3.6}$$

where $\tau_p(\omega_k)$ and $\tau_n(\omega_k)$ are the values of group delay function at the angular frequency ω_k due to pth pole and nth zero, respectively. The amount of

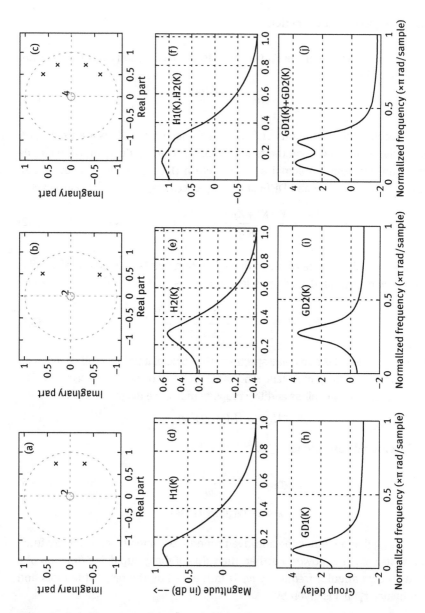

Figure 3.1: Resolving power of the group delay spectrum: z-plane, FT magnitude spectrum, group delay spectrum (a) pole inside the unit circle and (0.8, π/8), (b) pole inside the unit circle at (0.8, π/4) and (c) a pole at (0.8, π/8 and another pole at (0.8,π/4) inside the unit circle, (d), (e), (f) the corresponding magnitude spectra of these three systems, respectively, and (g), (h), (i) group delay spectra of these three systems.

influence of the adjacent poles on the group delay function of a given pole is proportional to the number of poles in a given system. In order to reduce the influence, the number of formants has to be reduced. This is accomplished by band-limiting the given speech signal. As the band of interest is low-frequency region, the signal is low pass filtered and formants are extracted using group delay-based formant extraction technique. Further, a group delay-based acoustic measure [6] is proposed to detect the presence of hypernasality in speech.

3.3.2.2 Selective pole modification-based technique for the detection of hypernasality

From our previous work, consistent effects of vowel nasalization in the low-frequency region were observed in hypernasal speech using group delay function. As a subsequent work, LP-based pole modification technique is used as a detection method for hypernasality. Unlike lower order LP analysis, which may not resolve two closely spaced formants, increasing the LP order may show spurious peaks in the low-frequency region for the normal speech. This leads to an ambiguity as to if a simple peak picking algorithm is sufficient for the detection of hypernasality [17]. On this note, this work concentrates on using a higher-order LP spectrum. The pole corresponding to strongest peak in the low-frequency region below the first oral formant is defocused and a new signal is resynthesized. In a spectrum, if a peak is corresponding to a true formant is suppressed, it would have a significant difference in the time-domain signal when resynthesized. This aspect is used for the detection of hypernasality.

For this study, the LP order is chosen to be 28 for the speech signal with a sampling rate of 16 kHz. As observed in our previous work based on group delay function, the nasal formant introduced due to hypernasality is stronger than the oral formant. In order to de-emphasize the influence of the nasal formant, the pole corresponding to the nasal formant with maximum radius below 300 Hz is defocused (reduce the radius) without altering the frequency. We empirically observed that if the radius is reduced by dividing the actual radius by a factor 1.15 suppresses the pole considerably without any significant effect on the rest of the frequency components of the signal. When these stronger components are de-emphasized, the other signal components that are suppressed due to the presence of strong nasal component are revealed clearly. Specifically, the signal components between consecutive excitations that are masked by the nasal resonance are seen clearly after selective pole-defocusing as illustrated in Figure 3.2. Apart from this, a considerable reduction in the signal strength is observed due to the suppression of the strong nasal formant.

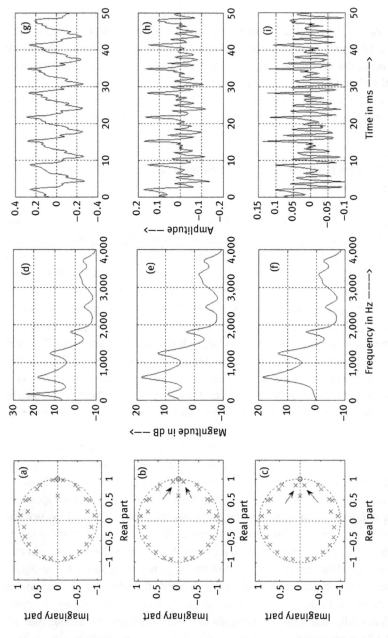

Figure 3.2: Effect of selective pole defocusing in the temporal and spectral domain: (a), (b) and (c) pole-zero plot of a hypernasal speaker showing pole defocusing, (d), (e) and (f) illustrate the corresponding effect in the spectral domain and (g), (h) and (i) show the effect of pole defocusing in the temporal domain.

Selective pole defocusing technique is applied on both hypernasal as well as normal speakers and further extended to the four dysarthric speakers who are declared as hypernasal by the group delay-based technique. For a normal speaker, as the peak below first formant is due to pitch harmonics, defocusing the peak does not have significant effect on the signal (refer to Figure 3.3). To check the similarity between the input and resynthesized signal after pole defocusing, normalized cross-correlation is computed. The correlation between the input and the modified signal is around 0.2 to 0.65 for hypernasal speakers, since the effect of defocusing the strongest pole (nasal formant) is significant in the time domain. In case of the normal speakers, the correlation is above 0.85 as illustrated in Figure 3.4. As normal speech has only pitch harmonic peak in the intended region and suppressing the corresponding pole has relatively less effect on the resynthesized signal resulting in high correlation.

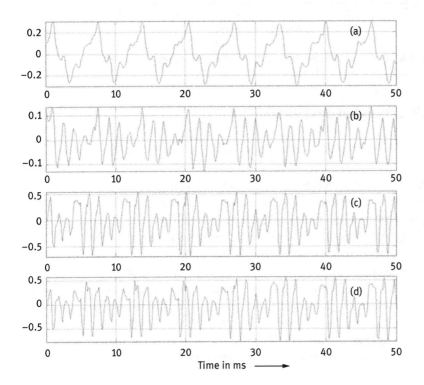

Figure 3.3: Effect of selective pole defocusing on hypernasal and normal speech in the temporal domain: hypernasal speech (a) before and (b) after pole defocusing; normal speech (c) before and (d) after pole defocusing.

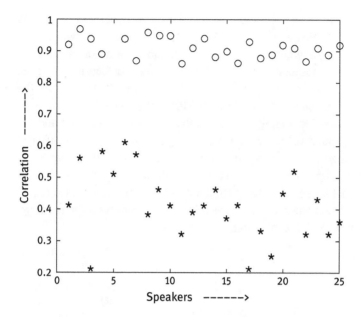

Figure 3.4: Correlation between input and resynthesized speech signals for hypernasal ("*" markers) and for normal speakers ("o" markers).

3.3.3 Laryngeal dysfunction

As a second step in our divide and conquer approach, the laryngeal dysfunction, if any, in dysarthric speech is analyzed. As discussed earlier laryngeal dysfunction in the speech signal is exhibited as jitter and shimmer variation. Figures 3.5 and 3.6 illustrate these variations. To carry out the analysis on laryngeal dysfunction, phonemes are extracted from the Nemours and the TIMIT speech corpora for the analysis. The speech rate was calculated for all the phonemes. Then acoustic analysis was performed on the vowels. This involved estimation of acoustic parameters and their comparison with the normal speakers. The latter consisted of normal speakers from the TIMIT speech corpus. Statistical, quantitative and graphical analyses were evaluated [18]. The reliability and validity of the underlying inferences projected are estimated by corroboration with FDA [8]. Correlation is calculated to quantitatively measure the success rate of the assessment.

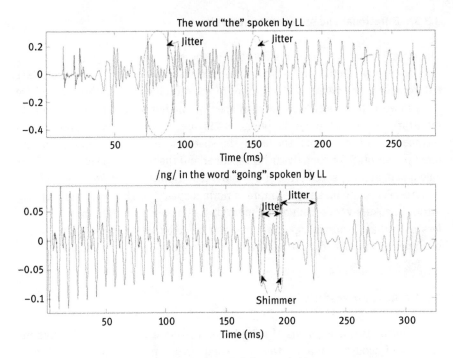

Figure 3.5: Laryngeal dysfunction reflected as pitch jitter and shimmer.

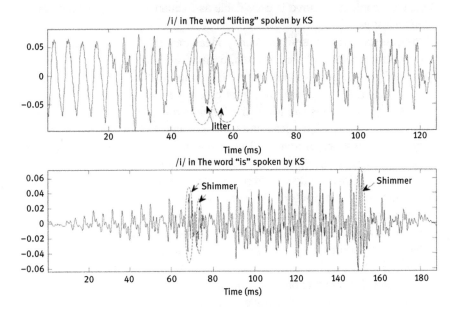

Figure 3.6: Illustration of jitter and shimmer variations in dysarthric speakers.

3.3.3.1 Durational and speech rate analysis

The duration of each phoneme for each of the speakers is computed using the time-aligned phonetic transcription available with the Nemours and the TIMIT database [19]. The means and variances of the duration of dysarthric speech are normalized with respect to the normal speech. The normalized mean duration of dysarthric speakers over all the phonemes is observed to be always greater than (~twice) that of the normal speakers. The speech rate (number of phonemes uttered per second) for each dysarthric speaker and the normal speakers (TIMIT) is calculated, and the speech rate of the individual dysarthric speaker is normalized with respect to the mean speech rate of normal speakers. The speech rate of dysarthric speakers varies from a minimum of 1.5 times to a maximum of 2.8 times less than that of normal speech. This may be reflected in the variation in the pitch period and hence an acoustic analysis is conducted to verify this.

3.3.3.2 Acoustic analysis

Fragmented speech segments of voiced sounds were extracted from the corpora. A range of acoustic variables were extracted. These included pitch period, variance in pitch period, pitch variation index (PVI), jitter and shimmer. The speaker information content procured is classifiable as frequency parameters and perturbation measures:

(a) **Frequency parameters:** Frequency measures give information regarding dysarthric speaker's habitual pitch, optimal pitch and pitch range, the degree of pitch steadiness and any pitch alterations during speech. Alteration in laryngeal function resulting from changes in vocal fold elasticity, stiffness, length or mass can affect the fundamental frequency. Consequently, statistical pitch period (mean and variance) an d PVI were useful for identifying the presence of abnormal laryngeal function and monitoring laryngeal function over time. These were indicators of identification of the presence of either reduced or excessively variable pitch use.

(b) **Perturbation parameters:** Acoustic perturbation measures reflect short- and long-term variability in the fundamental frequency, F_0, and amplitude. Specifically, measurement of short-term, cycle-to-cycle variability in F_0 is "jitter," while short-term variability in amplitude is "shimmer." These parameters are evaluated for all the dysarthric speakers available in the database to enhance speaker information detail. The parameters taken into consideration are as follows: jitter (local), jitter (relative average perturbation), shimmer (local) and shimmer (DDP). Here, these parameters are extracted using *Praat* [20, 21].

Both jitter and shimmer reflected high-frequency fluctuations; consequently, alterations in these values may be strongly associated with laryngeal tissue abnormalities, asymmetries in movement and fast-acting neuromuscular fluctuations. Specifically, the presence of jitter indicated variations in vibratory patterns of the vocal chords while shimmer indicated vocal fold instability. With respect to perceptual features, abnormalities in jitter and shimmer were closely associated with judgments of voice hoarseness, roughness and harshness.

(A) Observations
The acoustic variables were quantified and compared with FDA. Pearson's correlation coefficient was calculated to establish the degree of corroboration between this work and FDA.

(i) Speech rate analysis
The speech rate analysis proved to match with perceptual judgment to a superlative degree as is seen in Figure 3.7. While both exhibit the same pattern for the majority, speakers RL and SC alone show exception. Normal speakers converse at an average of 15 phonemes/second.

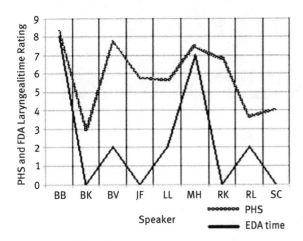

Figure 3.7: Comparison of phonemes/second and FDA laryngeal time rating for every dysarthric speaker.

(ii) Acoustic analysis

(a) **General patterns:** Certain common patterns in parameter variation were exhibited. Primarily, speech rate varied linearly with FDA laryngeal time rating. Similarly, PVI exhibited linear change with FDA laryngeal pitch rating while pitch period was inversely related.

(b) **Subject-specific analysis:** However, it was obvious that no standard patterns could be evolved to correlate acoustic parameters, speech rate and perceptual judgment (FDA ratings). The fact that each parameter showed individualistic variations, when corroborated with FDA, reinforced the fact the dysarthria may be caused by multiple subsystem disorders in specific combinations. Hence, a speaker-specific analysis was necessitated. To illustrate the speaker specific analysis, observations pertaining to few of the dysarthric speakers are narrated and the observations made are listed (refer to Figures 3.8, 3.9 and 3.10)

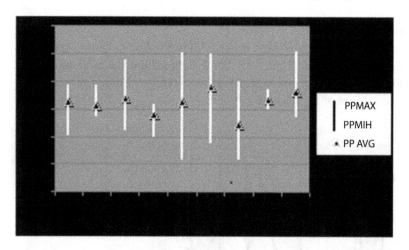

Figure 3.8: Maximum and minimum pitch periods with an average pitch period in ms for every dysarthric subject for voiced phoneme /aa/. (PP denotes pitch period.).

Speaker BB: Frequency parameters including pitch period, variance and PVI were of moderate in nature; his pitch range goes up to 7 ms. His jitter was mostly within the pathological threshold while his shimmer clustered around 5–10% indicating voice instability. He had highest a speech rate of 8 ph/s. Correspondingly laryngeal time and speech ratings by FDA were 8 with good overall assessment and word intelligibility 4, thus correlating with this work.

Speaker BK: BK's frequency parameters were in the normal range. The pitch period ranged from 5 ms to 7 ms. Variance in pitch period was extremely low,

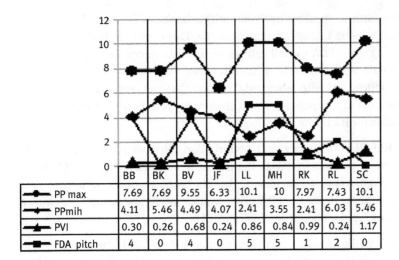

	BB	BK	BV	JF	LL	MH	RK	RL	SC
PP max	7.69	7.69	9.55	6.33	10.1	10	7.97	7.43	10.1
PPmih	4.11	5.46	4.49	4.07	2.41	3.55	2.41	6.03	5.46
PVI	0.30	0.26	0.68	0.24	0.86	0.84	0.99	0.24	1.17
FDA pitch	4	0	4	0	5	5	1	2	0

Figure 3.9: Comparison of pitch period (maximum), pitch period (minimum), pitch variation index and FDA laryngeal pitch rating for individual dysarthric subjects for the vowel /aa/.

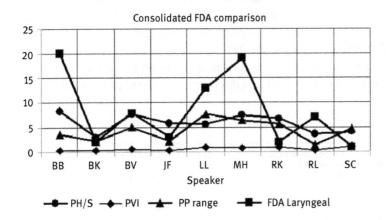

Figure 3.10: Consolidated analyses of speech rate and acoustic parameters (PVI and pitch period range) against consolidated FDA laryngeal rating (time, speech, pitch) for phoneme /aa/.

reflecting in the probability of vocal fold stiffness causing excess pitch steadiness and lack of modulation. Correspondingly, FDA laryngeal pitch rating was 0. His perturbation parameters were consistent. Jitter values varied around 1% while the shimmer went up to 11%, thus reflecting consistent laryngeal tissue abnormalities, symmetric movement and fast-acting neuromuscular fluctuations. Vocal folds could hence be relatively stable. However, BK was the slowest of the 10 dysarthric speakers available in the database. His speech rate was a meager 3 ph/s being totally incomprehensible. This correlated with FDA intelligibility rating of 0. FDA overall ratings were all extremely low. However, his jaw and palate were moderate in speech. Also, FDA summary states that the speaker produced /p/ with labiodental contact that can lead to slight build-up of pressure, lots of compensatory articulation. It is on these premises that perceptual rating for BK is low.

Speaker RL: RL's frequency and perturbation parameters were moderately good. His pitch period range was the lowest (6ms), indicating possible voice stiffness, roughness and hoarseness. He was found to be an extremely slow speaker at 4 ph/s. However, FDA gives a moderately low assessment with a time rating of 2 and moderate intelligibility of 4.

Speaker SC: While SC was a very slow speaker at 4 ph/s, his frequency parameters (PVI, variance and pitch period range) exhibited a lot of variations. Jitter was centered around 1% while his shimmer indicated chronic voice instability and laryngeal tissue abnormalities ranging up to 30%. This correlated with FDA assessment that indicated chronic dysarthria with intelligibility 1. Vowels may be of special difficulty to him, confirmed by FDA summary stating that lot of tasks confounded by lack of air support. SC compensates for many deficiencies using alternate means of production.

Pearson's correlation coefficient is deduced between FDA laryngeal assessment and the proposed assessment method and is tabulated in Table 3.1.

Table 3.1: Correlation between FDA parameters and acoustic parameters.

FDA parameter	Assessment parameter	Correlation coefficient
Laryngeal time	Speech rate	0.6425
Laryngeal (total)	Speech rate	0.6446
Laryngeal pitch	Pitch period range	0.6184
Laryngeal pitch	PP maximum	0.5187
Laryngeal pitch	PP average	0.4605

Total = pitch + time + in speech

Thus, speech rate and acoustic analysis performed on dysarthric data predicted the degree of severity and deduced the causes. The success rate was established by reasonably high correlation coefficients obtained by comparing against the standard perceptual judgment modality, FDA.

Although the described technique used Praat [21] for computations of all the parameters, any pitch estimation technique, like cepstrum-based, simplified inverse filtering and tracking-based, harmonic product spectrum-based [22], can be used for finding out the presence and severity of jitter and shimmer. However, the detection and, especially, the severity can be well estimated using glottal closure instants. Several successful algorithms are available in the literature, for example, average group delay-based [23], DYPSA [24], zero-frequency filter-based [25], to name a few. These techniques provide the locations of significant instants of excitation. Since the locations of excitations are known, the jitter and shimmer can be easily estimated with better accuracy.

To further analyze the effect of multisystem dysregulation, the dysfunctions present in articulatory subsystem is carried out using hidden Markov model-based speech recognition system as discussed below.

3.3.4 Articulatory dysfunction – analysis based on speech recognition system

The inability of dysarthric speakers to utter a subset of phonemes is directly reflected in the acoustic characteristics of the corresponding phonemes as deviations from their normal/healthy counterparts. Based on these factors, researchers have focused on utilizing automatic speech recognition (ASR) system either to improve intelligibility of dysarthric speech in order to develop assistive device [26–28] for the dysarthric speakers or for the assessment of dysarthric speech [29–32]. If an ASR is trained with normal speaker's speech data and tested with dysarthric speech data, it may clearly indicate the speech impairment rating. Using a word-level template-matching technique or a speech recognition system [31] to capture these acoustic variations is well established in the literature. Even though these techniques have been successfully used to quantify the intelligibility of dysarthric speech, they fail to indicate the inactive articulators. In an isolated word-based speech recognition system or in a distance measure-based system [30–33], the vocabulary should be carefully selected to derive information about the problems associated with the subsystems of speech. The recognition has to be performed in a closed set and the set of words should have only few different phonemes. Apart from this, in an isolated-word speech recognition system, the speakers are asked to speak the words in isolation. In case of mild or moderate dysarthric speakers if they speak words in isolation, they may produce it more clearly than in continuous speech, which may not give a clear picture about their problems in the speech subsystems.

A phoneme-based speech recognition system may be used to derive the information about the malfunctioning of speech subsystems. However, if a phoneme-based speech recognition system is used to analyze the continuous utterances of a dysarthric speaker, due to the variation in speech rate between normal (train data) and dysarthric speakers (test data), the performance of the system may be very poor. Here, we suggest and analyze the performance of an automatic, isolated-style phoneme recognition system for the assessment of dysarthric speech. Only a set of articulators are involved during the production of a phoneme or a sequence of phonemes. If a particular articulator does not function properly, the corresponding set of phonemes involving this articulator is expected to have a considerable deviation in acoustics when compared to that of the reference phonemes (from normal speech).The malfunctioning articulators can be pinpointed by grouping the set of phonemes based on their place of articulation. The error analysis can also be performed at the place-of-articulation level. Based on this, using phoneme as a subword unit, isolated-style phoneme recognition system is built for the assessment of articulatory subsystem of speech.

However, due to acoustic similarities between phonemes, the system may confuse a phoneme with other phonemes. In general, these confusions are more severe between the phonemes that have the same place of articulation. The performance of a system after phoneme grouping, based on place of articulation, for the assessment of subsystems of dysarthric speech, is analyzed and discussed further. The major advantages of such techniques are (a) vocabulary independence, (b) possibility of intelligibility analysis, as well as subsystems analysis and (c) very less manual intervention. As discussed earlier, the performance of a speech recognition system, trained with normal speech data (reference data) and tested with dysarthric speech data, may give a clear indication of the severity or intelligibility of the dysarthric speech. For this work, the speech data from the TIMIT speech corpus is considered as the reference data. The Nemours database of dysarthric speech is used for the assessment of dysarthria.

Here, the Nemours database of dysarthric speech is used for assessment and the TIMIT speech corpus is used for training. When a speech recognition system is used to analyze the malfunctioning of subsystems of speech, the recognition errors of such system are due to (a) acoustic similarity between phonemes and (b) due to inactive articulators. Theoretically, if the errors due to case (a) are completely removed, then the recognition performance of the resultant system will clearly indicate the problems associated with the subsystems alone. However, as these errors are unavoidable due to less amount of training data, the errors due to acoustic similarity and due to inactive articulators should be differentiated. In this work, these two errors are differentiated using product of likelihood Gaussian-based acoustic analysis. To reduce the errors due to acoustic similarity to a greater extent, context-dependent triphone models are used for training. The context-dependent (CD) triphone models capture the coarticulation effects on a phoneme better than the context-independent (CI) model and if a speech recognition system is trained with CD models the error due to acoustic similarity can be avoided to a greater extent. Further, the confusions across phonemes within the same place of articulation is considered as correct recognition as we are interested whether the speaker is able to use the articulator correctly and utter the phonemes corresponding to that place of articulation or not and rather than the manner of articulation.

(A) Experimental setup

The feature used for building the hidden Markov model-based speech recognition system using HTK [34] is the 39-dimensional Mel frequency cepstral coefficients with 13 static and 26 dynamic coefficients. Since the train and test data are different (the TIMIT and the Nemours), the feature vectors are cepstral mean subtracted to compensate the channel variation. Since the interest is shown only on finding out the acoustic-similarity of a test phoneme of a dysarthric speaker with the normal speakers' phoneme model, the decision-metric used is the acoustic log likelihood of the phonemes for the given models. To train context-dependent triphone models, the contexts are derived from the Nemours corpus. As the system is trained with the TIMIT corpus and tested with the Nemours corpus common triphones among these two corpora have to be located. It is found that there are 231 word-internal triphones available in the dysarthric speech, and only 128 triphones have more than 10 examples in the TIMIT speech corpus. All the triphones are trained with three states and eight-mixture components per state after mixture splitting. The performance of the context-dependent (triphone)-based speech recognition for each class of place of articulation [35] is compared with the respective sub-groups of the Frenchay dysarthria perceptual assessment.

It is observed that the performance of the system for each place of articulation is very well correlated with the FDA scores and the correlation scores are listed in Table 3.2. The overall performance for all the phonemes is used to corroborate with the intelligibility of the respective dysarthric speaker and correlated with word and sentence intelligibility in FDA. All the assessments made are speaker dependent as the effect of dysarthria is different for each of the dysarthric speaker. For instance, the speaker BB is able to utter all bilabial sounds correctly irrespective of the position of sound units (start, middle or end of the word) as observed in the recognition performance as well as in the perceptual

Table 3.2: Comparison of phone classes and FDA subgroups.

S. no	Place of articulation	Phones	Subgroups of FDA	Correlation coefficients
1	Bilabial	/b/, /p/, /m/	Lips in speech Lips seal Lips at rest Jaw in speech	0.8293
2	Dental / labiodental	/th/, /dh/, /f/, /v/	Lips at speech Lips spread Jaw in speech	0.8381
3	Alveolar	/t/, /d/	Tongue elevation Lips in speech Jaw in speech Palate maintenance Palate in speech	0.9225
4	Palatal	/ch/, /sh/	Jaw in speech Palate maintenance Palate in speech Tongue elevation	0.9245
5	Velar	/k/, /g/, /ng/	Tongue elevation Tongue in speech Jaw in speech	0.8709
6	Intelligibility	All sounds	Word and sentence intelligibility scores	0.9321

test. This is also correlated with his FDA scores in lips in speech and lips seal with a high score of each 7 out of 8. Similarly the recognition performance of the dental phones also correlated with the FDA scores in all three subgroups as mentioned in Table 3.2 with a high score of 7 each. However, the speaker always replaced the labiodental phones /f/ and /v/ by /p/. That is, the word /faith/ and /fife/ are uttered as /paith/ and /pipe/, respectively. Product of likelihood Gaussian [36] is used as a metric to find the acoustic similarity between the two phones uttered by the speaker. This metric can also be interpreted to quantitatively verify whether the speaker has actually mispronounced the phone /f/ as /p/. From the PoG analysis it is observed that the dysarthric speaker did pronounce /f/ as /p/. A detailed description on PoG analysis is given in [37, 38]

3.4 Intelligibility modification using speaker adaptation

As discussed in the previous sections, dysarthric speakers suffer from multiple dysfunctions that affect the intelligibility of the uttered speech. The various signal-processing techniques and speech recognition systems help us in assessing the dysfunctions, which may help the speech therapists to quantitatively diagnose the problems associated with the subsystem.

Dysarthric speakers' life style can be improved further by developing a text-to-speech synthesis system, which speaks in their own voice without the errors they usually make. Such a system mandates text as an input. However, a dysarthric speaker affected by quadriplegia or diplegia that affects their hand movement as well will make the problem more difficult. In such a case, the dysarthric speaker may use his own speech as an input to the speech recognition system, which will convert it to text and the resultant text may be used for re-synthesizing speech. However, as we know their speech may not be intelligible due to their problems, which may lead to poor recognition performance. To make use of speech as an input, let us utilize the observations made during the assessment of the speech disorder. From the previous acoustic analyses, the list of phones a particular speaker mispronounces or replaces is identified. This will help us in creating a speaker-specific dictionary with a correct pronunciation with respect to a particular dysarthric speaker and train a speech recognition system. Therefore, by combining the evidence derived from the assessment system and a speaker-dependent speech recognition system, appropriate text is provided as an input to a text-to-speech synthesizer.

For this study, the Nemours database is used for building a recognizer and a synthesizer for each of the dysarthric speaker in the database. This database has speech data from 10 dysarthric speakers uttering 74 sentences each, containing 110 unique words uttered two times by each of the dysarthric speaker. Due to data sparsity, as a proof of concept, in this work, context-independent monophone models for all the phones that are common to TIMIT speech corpus and Nemours database of dysarthric speech are trained using three-state and varying number of mixture components. The number of mixture-components per state for each phoneme is chosen based on the number of examples. The above-trained monophone HMMs are adapted to 10 dysarthric speakers of the Nemours dysarthric speech corpus individually using maximum a posteriori probability (MAP) adaptation technique leading to 10 dysarthric speaker-dependent ASR systems.

A speaker-specific pronunciation dictionary is created for each dysarthric speaker based on the list of articulatory errors estimated using acoustic likelihood analysis. The articulatory errors are considered as pronunciation variations. That is, a word with mispronounced phones can directly be mapped to the actual word itself.

Before we proceed to synthesizing speech, as we want our synthetic speech carrying the identity of the dysarthric speaker himself, the synthesis system has to be trained for the intended dysarthric speaker. However, to build a high-quality, HMM-based speech synthesis system (HTS) [39, 40] at least 1 h of speech data is required. However, the speech data available in the Nemours corpus for each speaker is very limited (just few minutes). At this juncture a speaker adaptation-based speech synthesis system (refer to Figure 3.11) will be an appropriate choice.

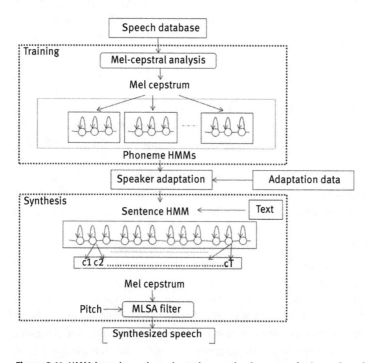

Figure 3.11: HMM-based speaker adaptation synthesis system (redrawn from [41]).

A text-to-speech synthesis system converts the text in any given language to the corresponding speech utterance. Techniques of adaptation such as MAP and maximum likelihood linear regression can be applied to HTS, trained with multiple speakers' data to obtain speaker-specific synthesis system with a small amount of speech data in the desired speaker's voice. It involves training context-dependent phoneme models, and concatenating models, specific to a given text, to synthesize speech. This technique, being a statistical parametric approach, produces good quality synthetic speech with minimum amount of data. Further, it is flexible in the sense that the model parameters can be manipulated to introduce any required variation. In this regard, the models can be adapted to take up a desired speaker's characteristics, with minimal amount of data from that speaker.

In the training phase, the spectral parameters, namely, 35-dimensional Mel-generalized cepstral coefficients (mgc) and their first and second derivatives, and the excitation parameters, namely, the log-fundamental frequency and its first and second derivatives, are extracted from CMU Arctic (4 male speakers) database. These parameters are used to train a speaker-independent (average voice), context-dependent (pentaphone) HMMs. In the adaptation phase, spectral and excitation parameters are extracted from a small amount (~2 min) of a dysarthric (target) speaker's data. These are used to adapt the HMMs trained in the training phase to those of the target speaker. Here, constrained maximum likelihood linear regression-based technique is used for speaker adaptation [42–45]. At the synthesis phase, given a text, speech is generated in the dysarthric speaker's voice by extracting spectral and excitation parameters from the appropriate adapted models, using a Mel-logarithmic spectrum approximation filter. A perceptual test with a 5-point grading is conducted to verify whether the adapted speech sounds like dysarthric speaker. It is observed that the speaker identity of the dysarthric speaker with perceptual rating of 4 is obtained and the phone units that are not pronounced by the speaker or not available in his data during adaptation are now replaced by the phonemes with correct pronunciation, in his own voice, making the sentence more intelligible.

3.5 Summary

This chapter has focused on two main areas, namely, the assessment of dysarthric speech considering each speech subsystems independently and intelligibility modification of the moderately unintelligible dysarthric speech. The purpose of this chapter is to provide an idea of characteristics of dysarthric speech and possible temporal and spectral parameters that characterize the disordered speech. It is also shown how effectively speech technology can be utilized to support and enhance the speaking capability of speakers affected by a multisystem dysregulation of speech.

Acknowledgment: The authors extend their thanks to the authorities of All India Institute of Speech and Hearing (AIISH), Mysore, and SRMC, Chennai, India, The authors are thankful especially to Dr Savithri, AIISH, Mysore, and Dr Roopa Nagarajan, SRMC, Chennai, India, for extending their help in collecting hypernasal speech data.

References

[1]   Kent, R. D., Kent, J. F., Weismer, G., and Duffy, J. R. "What dysarthrias can tell us about the neural control of speech". Journal of Phonetics, 28, 273–302, 2000.
[2]   Kent, R. D. "Research on speech motor control and its disorders: A review and prospective". Journal of Communication Disorders, 33, 391–428, 2000.
[3]   Kent, R. D., Weismer, G., Kent, J. F., Vorperian, H.K., and Duffy, J. "Acoustic studies of dysarthric speech". Journal of Communication speech, 32, 141–186, 1999.
[4]   http://www.asha.org
[5]   Ray, D. Kent, Weismer, Gary., Kent, Jane F., Vorperian, Houri K., and Duffy, Joseph R. "Acoustic studies of dysarthric speech: Methods, progress and potential". Journal of Communication Disorders, 32(3), 141–186, May 1999.
[6]   Vijayalakshmi, P. "Detection of hypernasality and assessment of dysarthria using signal processing techniques", Ph.D dissertation, Biomedical Engineering Group, Indian Institute of Technology, Madras, 2007.
[7]   Pidal, X. M., Polikoff, J. M., Peters, S. M., Leonzio, J.E, and Bunnell, H. T. "The nemours database of dysarthric speech". in Proc. of Int. Conf. on Spoken Language Processing, Philadelphia, 1962–1965, 1999.
[8]   Enderby, P. "Frenchay dysarthric assessment". British Journal of Disorders and Communications, 15(3), 165–173, 1980.
[9]   Fisher, W. M., Doddington, G. R., and Marshal, K. M. "The DARPA speech recognition research database: Specifications and status". in Proc. of DARPA workshop on Speech Recognition, 93–99, 1986.
[10]   Kominek, J., and Black, A. "The CMU Arctic speech databases". in Proc. 5th ISCA speech synthesis workshop, Pittsburgh, USA, 223–224, 2004.
[11]   Dworkin, J. P., Marunick, M. T., and Stachler, R. J. "Neurogenic velopharyngeal incompetence: Multidisciplinary diagnosis and treatment". The Journal of Applied Research in Clinical Dentistry, 1(1), 35–46, 2004.
[12]   Vijayalakshmi, P., Reddy, M. R., and Douglas, O'Shaughnessy. "Acoustic analysis and detection of hypernasality using group delay function". – IEEE Trans on Biomedical Engineering, Apr, 54(4), 621–629, 2007.
[13]   Vijayalakshmi, P., and Reddy, M. R. "Analysis of hypernasality by synthesis," in Proc. of Int. Conf. on Spoken Language Processing, Jeju Island, South Korea, 525–528, 2004.
[14]   Murthy, H. A., and Gadde, V. "The modified group-delay function and its application to phoneme recognition. in Proc. of IEEE Int. Conf. Acoust., Speech, and Signal Processing, Apr., 68–71, 2003.
[15]   Murthy, H. A., and Yegnanarayana, B. "Formant extraction from minimum phase group delay function". Speech Communication, 10, Aug., 209–221, 1991.

[16] Hawkins, S., and Stevens, K. N. "Acoustic and perceptual correlates of the non-nasal-nasal distinction of vowels". Journal of Acoustic Society of Amer, 77, 1560–1574, 1985.

[17] Vijayalakshmi, P., Nagarajan, T., and Jayanthan, Ra. V. – "Selective pole modification-based technique for the analysis and detection of hypernasality". in Proc. of IEEE TENCON, Nov., 1–5, 2009.

[18] Surabhi, V., Vijayalakshmi, P., Steffina, Lily T., and Jayanthan, Ra. V. – "Assessment of laryngeal dysfunctions of dysarthric speakers". in Proc. of IEEE EMBC, Sep., 2908–2911, 2009.

[19] Vijayalakshmi, P., Nagarajan, T., and Reddy, M. R. "Assessment of articulatory and velopharyngeal sub-systems of dysarthric speech". Biomedical Soft Computing and Human Sciences, 14(2), 87–94, 2009.

[20] Boersma, Paul., and Weenink, David. (2014). Praat: Doing Phonetics by Computer [Computer program], Version 5(3),64, http://www.praat.org/

[21] Pascal, V. Lieshout. – "Praat, a basic introduction". – http://www.stanford.edu

[22] Rabiner, L. R., and Schaffer, R. W. "Digital processing of speech signals", Prentice Hall, 1978.

[23] Murthy, P. S., and Yegnanarayana, B. "Robustness of group-delay-based method for extraction of significant instants of excitation from speech signal". IEEE Transactions on Speech and Audio Processing, 7(6), 609–619,1999.

[24] Kounoudes, Anastasis, Naylor, P. A., and Brookes, M. "The DYPSA algorithm for estimation of glottal closure instants in voiced speech". IEEE International Conference on Acoustics, Speech, and Signal Processing, 349–353, 2002.

[25] Murty, K. S. R., Yegnanarayana, B., and Joseph, M. Anand. "Characterization of glottal activity from speech signals". IEEE Signal Processing Letters, 16(6),469–472, 2009.

[26] Deller, J. R., Hsu, D., and Ferrier, L. J. "Encouraging results in the automated recognition of cerebral palsy speech". IEEE Trans. Biomedical Engineering, 35(1), 218–220, 1998.

[27] Hosom, J. P., Kain, A. B., Mishra, T., Santen, J. P.H., Oken, M. F., and Staehely, J. "Intelligibility of modifications to dysarthric speech". in Proc. of IEEE Int. Conf. Acoust., Speech, and Signal Processing, 924–927, 2003.

[28] Sanders, E., Ruiter, M., and Beijer, L Strik. "Automatic recognition of Dutch dysarthric speech – a pilot study". in Proc. of Int. Conf. on Spoken Language Processing, Denver, 661–664, 2002.

[29] Carmichael, J., and Green, P. "Revisiting dysarthria assessment intelligibility metrics". in Proc. of Int Conf. on Spoken Language Processing, 485–488, 2004.

[30] Sy, B. K., and Horowitz, D. M. "A statistical causal model for the assessment of dysarthric speech and the utility of computer based speech recognition". IEEE Trans. on Biomedical Engineering, 40(12), 1282–1298, 1993.

[31] Pidal, X. M., Polikoff, J. B., and Bunnell, H. T. "An HMM-based phoneme recognizer applied to assessment of dysarthric speech". in EUROSPEECH, 1831–1834, 1997.

[32] Vijayalakshmi, P., Reddy, M. R., and Douglas, O'Shaughnessy. – "Assessment of articulatory sub-systems of dysarthric speech using an isolated-style speech recognition system" – in Proceedings of Int. Conf. on Spoken Language Processing (ICSLP), INTERSPEECH, Pittsburgh, Sep. 2006, 981–984.

[33] Gu, L., Harris, J. G., Shrivastav, R., and Sapienza, C. "Disordered speech evaluation using objective quality measures". in Proc. of IEEE Int. Conf. Acoustic Speech, and Signal Processing, 1, 321–324, 2005.

[34] Young, S., Evermann, G., Hain, T., Kershaw, D., Moore, G., Odell, J., Ollason, D., Povey, D., Valtchev, V., and Woodland, P. (2002) HTK Book, ver. 3.2.1, Cambridge University Engineering Department.

[35] Vijayalakshmi, P. and Reddy, M. R. "Assessment of dysarthric speech and an analysis on velopharyngeal incompetence". Proceedings of the 28th IEEE EMBS Annual International Conference, New York City, USA, Aug 30–Sep. 3, 2006.

[36] Nagarajan, T., and O'Shaughnessy, D. "Bias estimation and correction in a classifier using product of likelihood-Gaussians". in Acoustics, Speech and Signal Processing, 2007. ICASSP 2007. IEEE International Conference on, 3, Apr, 1061–1064, 2007.

[37] Dhanalakshmi, M., Mariya Celin, T.A., Nagarajan, T., and Vijayalakshmi, P. "Speech-Input Speech-Output communication for dysarthric speakers using HMM-based speech recognition and adaptive synthesis system". Circuits, Systems, and Signal Processing, 37(2),674–703, 2018.

[38] Dhanalakshmi, M., and P. Vijayalakshmi. "Intelligibility modification of dysarthric speech using HMM-based adaptive synthesis system." in proceedings of IEEE 2nd International Conference on Biomedical Engineering (ICoBE), 1–5, 2015.

[39] Zen, H., Tokuda, K., and Black, A. W. "Statistical parametric speech synthesis". Speech Communication, 51, 1039–1064, Nov. 2009

[40] Tokuda, K., Zen, H., and Black, A. W. "An HMM-Based speech synthesis system applied to English". in Proc. of IEEE Workshop on Speech Synthesis, 227–230, 2002.

[41] Masuko, T., Tokuda, K., Kobayashi, T., and Imai, S. "Voice characteristics conversion for hmm-based speech synthesis system". in Proceedings of IEEE Int. Conf. Acoust., Speech, and Signal Processing, 1611–1614, 1997.

[42] Leggetter, C. J., and Woodland, P. C. "Maximum likelihood linear regression for speaker adaptation of continuous density hidden Markov models". Computer Speech and Language, 9(2), 171–185, 1995.

[43] Ferras, M., Leung, C. C., Barras, C., and Gauvain, J.-L. "Constrained MLLR for speaker recognition". in Proceedings of IEEE Int. Conf. Acoust., Speech, and Signal Processing, 4, 53–56, 2007.

[44] Tamura, M., Masuko, T., Tokuda, K., and Kobayashi, T. "Speaker adaptation for HMM-based speech synthesis system using MLLR". – the third ESCA/COCOSDA Workshop on Speech Synthesis, 273–276, 1998.

[45] Gales, M.J.F. "Maximum likelihood linear transformations for HMM-based speech recognition". Computer Speech and Language, 12(2), 75–98, 1998.

Part II: **New approaches to speech reconstruction and enhancement via conversion of non-acoustic signals**

Seyed Omid Sadjadi, Sanjay A. Patil and John H.L. Hansen

4 Analysis and quality conversion of non-acoustic signals: the physiological microphone (PMIC)

Abstract: There are a number of scenarios where effective human-to-human speech communication is vital, yet either limited speech production capabilities due to pathology or in noisy environmental conditions limit intelligible information exchange and reduce overall quality. Traditionally, front-end speech enhancement techniques have been employed to alleviate the effect of environmental noise. Speech-processing normalization of speech under pathology has also been employed to increase quality of pathological speech. A recent alternative approach to deal with these scenarios is the use of nonacoustic sensors, which are essentially independent of the acoustic sound propagation characteristics of the environments where human communication is needed. The physiological microphone (PMIC), as a nonacoustic contact sensor, has been shown to be quite useful for speech systems under adverse noisy conditions. It also could provide an alternative signal capture mode for speech under vocal fold pathology. However, due to alternative pickup location and the nonacoustic principle of operation, captured signals appear muffled and metallic to the listener with variations to the speaker-dependent structure. To facilitate more robust and natural human-to-human speech communication, in this chapter we present a probabilistic transformation approach to improve the perceptual quality and intelligibility of PMIC speech by mapping the nonacoustic signal into the conventional close-talk acoustic microphone speech production space, as well as by minimizing distortions arising from alternative pickup location. Performance of the proposed approach is objectively evaluated based on five distinct measures. Moreover, for subjective performance assessment, a listening experiment is designed and conducted. Obtained results confirm that incorporating the probabilistic transformation yields significant improvement in overall PMIC speech quality and intelligibility. This solution offers an alternative to individuals with severe vocal fold pathology or areas where traditional speech production is not an option.

Keywords: nonacoustic microphone, PMIC, voice transformation, speech quality improvement, intelligibility improvement, MOS

Seyed Omid Sadjadi, Sanjay A. Patil, John H. L. Hansen, Center for Robust Speech Systems (CRSS), University of Texas at Dallas, USA

https://doi.org/10.1515/9781501501265-005

4.1 Introduction

Adverse noisy conditions can pose problems to both human–human and human–machine speech communications since they lead to reduced speech quality and intelligibility [1]. Even when intelligibility remains high, the degraded speech quality and loss of acoustic cue context causes an increase in the cognitive load of the listener, often referred to as *listener fatigue* [2, 3], but with limited quantitative methods for assessment.

To overcome these challenges and possible errors or loss in message transfer, some adaptations are naturally adopted by either the speaker or listener including increased vocal effort, hyperarticulation of the speech message in a *clear* or *shouted* manner (Lombard effect) [4, 5], repeating the message, augmenting acoustic decoding based on visual clues such as lip movements and other body gestures. Nevertheless, these are all useful when the time delay is not a critical issue and/or the communication is face-to-face. In certain situations, such as in speech communication among first responders for emergency/ medical scenarios (e.g., fire fighters), workers in a factory environment with heavy-metal manufacturing facilities, and aircraft pilots during take-off and landing[1] where speech is transmitted via a communication channel using electronic transducers and interaction response time is vital, the above-mentioned adaptations may not be helpful.

While situational or environmental factors could compromise speech quality and intelligibility to the level of requiring alternative nonacoustic speech capture, there is also the domain of vocal fold speech pathology that could benefit. Changes in speech production due to vocal fold pathology is a major focus in developing models and understanding speech production when physical or motor issues impair speech production. Past research has considered direct methods for estimating speech feature changes due to vocal fold cancer [6], as well as nonlinear speech operators (i.e., TEO-CB-AutoEnv) for detection and classification of speech pathology [7]. Automatic detection of other vocal impairments have also been considered based on short-term cepstral parameters [8], hyper-nasality detection based on nonlinear TEO operators [9] and quality assessment using GMMs for vocal pathologies [10]. However, assessing and detecting/classifying vocal fold or excitation-based pathologies are only part of the challenge in this domain. The ability to transform the resulting pathology speech into an acoustic speech signal with a consistent degree of quality and

---

1 The noise exposure in aircraft communications is typically greater than 90 dBA-SPL in these environments.

intelligibility represents the subsequent challenge. The use of nonacoustic sensors, capable of capturing low-quality/intelligible speech, followed by transformational algorithms capable of morphing the pathological speech into a signal with high quality is a key area for improving human-to-human communications when a subject has a vocal fold pathology.

The basic question at this point could be exploring ways to enhance speech quality of such nonacoustic captured speech. Here, the use of electronic devices with traditional speech enhancement algorithms for improving speech quality and intelligibility has been employed for several decades. These enhancement techniques are noise suppression algorithms, feature enhancement approaches and so on (for a review see Loizou [11] or Hansen [12]). While such approaches appear to improve degraded signal quality, they may introduce undesirable processing distortion (e.g., artifacts), and for extremely low signal-to-noise ratio (SNR) scenarios, in which the signal is completely obscured in noise, provide limited or generally no improvement in intelligibility.

A recent alternative approach to deal with these adverse scenarios is the utilization of nonacoustic contact sensors, in order to acquire the speech signal as close to the speech production apparatus as possible. Due to their nonacoustic principles of operation, they are relatively immune to acoustic characteristics (e.g., background noise and room reverberation) of the environments under which they are operating. Although basically different in signal capture mechanisms, such sensors record the glottal activity that occurs during phonation. The physiological microphone (PMIC) [13], throat microphone (TM) [14] and nonaudible murmur microphone [15–17] capture the speech signal through skin/soft-tissue vibration pickup, while for the bone-conduction microphone (BCM) [18] the signal is acquired by means of bone vibration. The glottal electromagnetic micropower sensor (GEMS) [19] and tuned electromagnetic resonating collar sensor [20] are other examples of nonacoustic sensors, which measure the movement of articulators during voiced speech through recording of the electromagnetic waves reflecting off the speech production apparatus.

Nonacoustic sensors have all been successfully applied to single and multisensor noise-robust speech systems. Subramanya et al. [21] exploited the BCM along with a close-talk microphone (CTM) in a multisensory speech enhancement framework. Quatieri et al. [22] considered the use of GEMS, PMIC and BCM in low-rate speech coding to improve the intelligibility in harsh acoustic environments. Both the PMIC and TM have been shown to be quite effective for automatic speech recognition (ASR) tasks, especially at relatively low SNR situations where the use of conventional microphone technology fails [13, 23, 24]. The PMIC has also been shown to be more successful than a CTM for speaker assessment and verification tasks under noisy stressful conditions [25].

Although the nonacoustic sensors are robust against acoustic disturbances, due to alternative pickup location and nonacoustic principles of operation, the captured signals do not necessarily reflect lip radiation and direct vocal tract resonance structure. In addition, the spectral content of the captured signal is limited by frequency characteristics of the body tissues. Furthermore, because these sensors capture a surface vibration wave produced by speech organs during voiced speech, the unvoiced portion of speech is weakly presented in their signals. All these give rise to reduced quality of the speech acquired from nonacoustic sensors, when compared to that acquired from a conventional CTM. As a consequence, along with these sensors an additional postprocessing stage would be useful to enhance overall quality and intelligibility of their signals before delivery to listeners, which could be either machines or humans.

Different techniques have been reported in the literature as the postprocessing stage to minimize the distortions arising from the alternative signal pickup location for different types of nonacoustic sensors [23, 26–28]. Although under different names and formulations, such techniques share the same concept: mapping the nonacoustic signal into the conventional speech production space. This is realized by mapping the speech spectra from the nonacoustic sensor to the CTM response.

Following this concept, in this chapter[2] we focus on use of the PMIC for noise-robust signal capture and present a standard probabilistic transformation approach as the postprocessing stage. The transformation approach closely follows the filtering technique described in Neumeyer and Weintraub [29]. Our objective is to improve overall quality and intelligibility of speech signals acquired by the PMIC for human listeners, in order to reduce their *listening effort* thereby facilitating a more robust human-to-human speech communication. It is worth mentioning here that, except for the methods proposed in Tran et al. [26] and Toda et al. [16, 17] that were subjectively assessed to measure the quality as well as intelligibility of estimated speech signals, none of the aforementioned techniques were evaluated with human listeners, but rather they have been evaluated using either objective measures [27], or in terms of word error rates (WER) obtained from ASR tasks [23, 28].

Assuming the linear source-filter speech production model, our approach involves mapping the filter parameters (i.e., the spectral envelope and gain) of the PMIC signal to estimate the CTM speech. Line spectral frequencies (LSF) [30] are employed to represent the spectral envelope. Both the PMIC and CTM acoustic spaces are modeled via codebooks obtained from vector quantization (VQ) using only a small amount of parallel training data, and the mapping is

---

2 A preliminary portion of this study was presented at INTERSPEECH-2010 [50].

performed in a piecewise linear manner. Instead of performing a hard-decision (i.e., one-to-one mapping) as implemented in a traditional VQ-based transform (e.g., Abe et al. [31]), we incorporate a probabilistic soft-decision strategy within the VQ framework, which helps reduce processing artifacts associated with the mapping process. The parameters of the mapping function are determined by solving a linear least square (LS) estimation problem in each VQ cluster. The effectiveness of the proposed method in removing the metallic and muffled nature prevalent in PMIC signals is evaluated based on both objective quality measures and formal listening tests. As a visual measure, spectrotemporal characteristics of the PMIC signal, which are correlated with quality, before and after the mapping are also studied and compared against that of the CTM.

4.2 Acoustic analysis of PMIC signals

The PMIC is a contact sensor capable of capturing the speech-related signals through skin vibrations [32]. It is a nonacoustic device having a piezo-crystal sensor encapsulated in a gel-filled cavity (Figure 4.1). The PMIC is typically strapped to any part of the body that allows faithful pickup of skin vibrations during the occurrence of speech resonances. Studies have considered speech collection with the device strapped on the forehead or wrist, but usually it is strapped around the throat near the carotid and thyroid cartilage [32], as shown in Figure 4.1. When the PMIC is strapped around the throat (as is the case for this study), skin vibrations during speech phonation are picked up by the gel-filled cavity and then transferred through the gel substance to the sensor. Due to this particular impedance matching property and also the material property of high air-borne acoustic coupling index, the PMIC signal is relatively immune to environmental acoustic noise.

Figure 4.1: The PMIC sensor (right) and its placement around the throat region identified with a white circle (left).

Figure 4.2 shows sample spectrograms and their corresponding time waveforms for the PMIC and CTM sensors, obtained simultaneously so that direct
temporal comparisons can be made. As described earlier, the PMIC is strapped
around the neck; hence, it is close to the glottal excitation source (the vocal
folds) and acquires speech via skin/throat vibrations. This results in a muffled-
like sound as compared to the CTM, which captures speech via sound pressure
acoustic waves at the lip region. The PMIC signal is suggested to be a fre-
quency-limited version of the CTM signal (see Figure 4.2); thus, it is perceptually less intelligible than the CTM signal. For some speech segments, the PMIC
might even sound annoying and require long-term *listening effort* by the
listener.

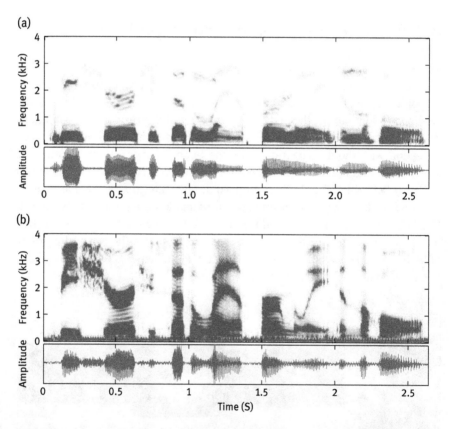

Figure 4.2: Spectrotemporal comparison between (a) PMIC and (b) CTM signals for the
utterance "be sure to set the lamp firmly in the hole."

It can be seen from Figure 4.2 that the PMIC signal possesses predominantly low-frequency content (less than 1 kHz), although weak activity replicating the CTM spectrogram can also be observed between 1 kHz and 3 kHz. Moreover, high-frequency consonant structure such as fricatives, stops or affricates are severely suppressed in the PMIC spectrogram compared to that of the CTM.

There exists a loss in the intelligibility and speech quality for PMIC signal recordings based on the following observations:

1) A study on the significance of different bands of frequency toward speech intelligibility reveals that a maximum contribution is from frequency bands near or above 2.5 kHz [33]. However, the PMIC shows a weak frequency response in this region.
2) The formant pattern continuity in a vowel is necessary for correct perception of the vowel [34]. Because of the low-pass frequency response nature of the PMIC, a certain loss is expected in vowel perception, as well as how coarticulation effects of vowels contribute to correct consonant recognition.
3) Because formant broadening (or flattening) impacts perceived quality, omission or relocation of the formants (as with the PMIC) will deteriorate the overall perceived speech quality, based on knowledge of how these factors alter human perception [1, 34].
4) It has a fast roll-off for frequencies above 2.7 kHz. This is due to the fact that (i) the body tissue acts as a low-pass filter and attenuates spectral energy in higher frequency regions and (ii) the PMIC basically captures the glottal activity that occurs during phonation; therefore, the unvoiced portion of speech is weakly presented in its signals.

Nevertheless, the PMIC speech is still intelligible for the following reasons:

1) Strong frequency components that are present below 1500 Hz exist in the PMIC signal; hence, the temporal structure is preserved. Comprehending the speech content is possible albeit with careful listening (e.g., increased *listening effort*).
2) Knowledge of the location of spectral peaks (i.e., formants) is necessary since spectral details within the region of formant peaks are more likely to stand out and be preserved in the presence of background noise. Considering the fact that at least two strong formats exist within the PMIC frequency response, a certain level of intelligibility is expected. Therefore, it is expected that the PMIC can preserve a measureable portion of the vocalic identity [1].
3) Auditory frequency selectivity is sharpest at low frequencies and diminishes with an increase in frequency [35].

Figure 4.3 illustrates average linear prediction (LP) spectra and corresponding LSFs of the (a) front vowel /i/ as in "evening," (b) back vowel /u/ as in "often," (c) nasal consonant /n/ as in "in" and (d) fricative consonant /sh/ as in "she," for PMIC and CTM recordings, respectively.

For the front vowel /i/, the PMIC LP spectrum shows movement of the third and fourth formants toward each other, resulting in broadening of their band-widths (Figure 4.3(a)). The second formant is also broadened to some extent. It is known from the study by Nakatani and McDermott [34] that pitch and for-mant shifts can impact speech quality, and it can be seen from the figure that there is a spectral formant change that occurs for /i/; hence, it is expected that the PMIC will be less intelligible.

It can also be seen from Figure 4.3(b) that the third formant of the PMIC back vowel /u/ is missing. Moreover, the PMIC strengthens the fourth formant and weakens the spectral peaks near 1.25 kHz, which causes a change in the spectral tilt. According to Assmann and Summerfield [1], spectral tilt is impor-tant in perception of the different phonemes. Therefore, based on changes in spectral tilt, it is expected that the PMIC will be less perceptible than the CTM in this case.

An interesting observation can be made from Figure 4.3(c) in which the LP spectrum of the PMIC nasal consonant /n/ is more like a front vowel than a nasal, and is quite different from that of the CTM. The nasal sounds are pro-duced by a glottal excitation waveform with an open nasal cavity and closed oral cavity. Therefore, the CTM, which operates based on the acoustic air pres-sure, captures a weak energy profile in the high-frequency bands. Because of the acoustic coupling between the pharyngeal and nasal cavities for nasals, the PMIC records the resulting spectral resonances in the overall vocal system, making the nasal spectra closer to a vowel.

Figure 4.3(d) shows that the LP spectrum for the fricative consonant /sh/ has neither frequencies below 1.5 kHz nor strong frequency content around or beyond 2.5 kHz. Because of the strong spectral structure in the 3.5 kHz region in the sound /sh/, the PMIC is able to capture the sound faithfully, even though the third formant at 2.6 kHz is shifted closer to the fourth formant. These obser-vations point to the lack of speech quality and intelligibility in the PMIC speech, which reflects our prime motivation to formulate an algorithm to im-prove these aspects. The proposed approach should be able to reconstruct, and if possible strengthen, weak formant peaks, thus improving overall vocalic qualities. The transformation from PMIC to CTM speech should satisfy the fol-lowing constraints: (i) it should limit perceptual confusion arising in PMIC speech due to its inherent band-limited response and (ii) it should minimize

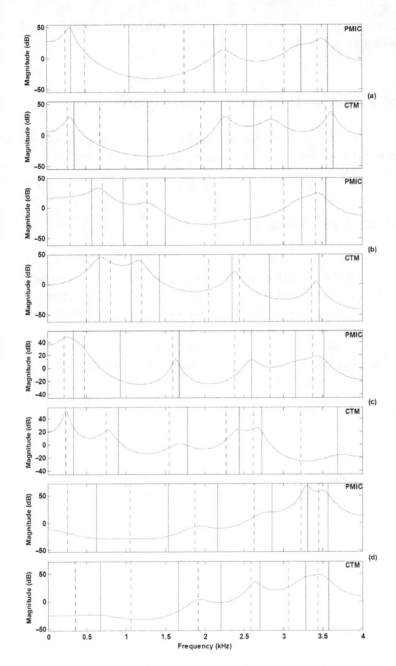

Figure 4.3: The averaged LP spectra and corresponding LSFs for the (a) front vowel /i/, (b) back vowel /u/, (c) nasal consonant /n/ and (d) fricative consonant /sh/.

any potential distortion that may arise during the transformation process. Such a transformation is proposed in the next section.

4.3 Probabilistic transformation algorithm

In this section, we first formulate the probabilistic transformation algorithm, and then consider implementation for mapping the PMIC signal into the conventional CTM speech response.

4.3.1 Problem formulation

Assuming that parallel signal data are available from the PMIC and CTM, our goal is to minimize the mismatch between spectral responses of the two sensors represented by the spectral envelopes extracted from their corresponding recordings. Our expectation is that in this manner the PMIC signal quality will move to that perceived from a comparable CTM signal. More precisely, let $\{\mathbf{x}_t, t = 1, \ldots, N\}$ and $\{\mathbf{y}_t, t = 1, \ldots, N\}$ denote two sets of corresponding PMIC and CTM n-dimensional vectors of LSF, representing the spectral envelopes (N is the total number of training frames). A linear mapping function $\Phi(\cdot)$ that transforms each vector of the set $\{\mathbf{x}_t\}$ into its counterpart in the set $\{\mathbf{y}_t\}$ is defined as

$$\Phi(\mathbf{x}_t) = \mathbf{A}^T \mathbf{x}_t + \mathbf{b}, \tag{4.1}$$

where $\mathbf{A}$ is an $n \times n$ matrix, and $\mathbf{b}$ is an $n \times 1$ additive term, both of which are to be estimated. Rearranging (eq. (4.1)) into a single matrix equation, we obtain

$$\Phi(\mathbf{x}_t) = \mathbf{H}^T \widetilde{\mathbf{x}}_t, \tag{4.2}$$

where

$$\widetilde{\mathbf{x}}_t = \begin{bmatrix} \mathbf{x}_t^T & \vdots & 1 \end{bmatrix}^T, \quad \mathbf{H} = \begin{bmatrix} \mathbf{A}^T & \vdots & \mathbf{b} \end{bmatrix}.$$

Now, $\Phi(\cdot)$ is entirely defined by the $(n+1) \times n$ transformation matrix $\mathbf{H}$. This matrix is computed by a linear LS optimization based on the parallel training data to minimize the sum of squared transformation errors,

$$\epsilon = \sum_{t=1}^{N} \|\mathbf{y}_t - \Phi(\mathbf{x}_t)\|^2 \tag{4.3}$$

The optimal solution for the matrix $\mathbf{H}$ is given by

$$\mathbf{H}_{\text{opt}} = \mathbf{R}_{\mathbf{xx}}^{-1}\,\mathbf{r}_{\mathbf{yx}}, \tag{4.4}$$

in which $\mathbf{R}_{\mathbf{xx}} = \Sigma_{t=1}^{N}\mathbf{x}_t\widetilde{\mathbf{x}}_t^T$ is the autocorrelation matrix of the vectors in the set $\{\mathbf{x}_t\}$, and $\mathbf{r}_{\mathbf{yx}} = \Sigma_{t=1}^{N}\mathbf{y}_t\widetilde{\mathbf{x}}_t^T$ is the cross-correlation matrix of vectors in the two sets.

To gain a better interpolation, thus compensating for possible losses in acoustic information discussed in Section 2, the feature vector $\widetilde{\mathbf{x}}_t$ can be modified to have the form $\widetilde{\mathbf{X}}_t = \begin{bmatrix} \mathbf{x}_{t-k}^T & \cdots & \mathbf{x}_t^T & \cdots & \mathbf{x}_{t+k}^T & 1 \end{bmatrix}^T$, with k being the number of frames in the neighborhood of the current frame t [16, 17, 29]. Moreover, in order for $\Phi(\cdot)$ to perform well across different phoneme classes within speech, it should take into account the classification of the PMIC acoustic space. Accordingly, the general transformation matrix $\mathbf{H}$ turns into a class specific version, and the transformation can then be performed piecewise linearly. Classification can be realized through a VQ framework; however, the discrete nature (i.e., one-to-one mapping) inherent in VQ may hurt the transformation quality. As a remedy, we fit a GMM onto the codebook obtained from the VQ and incorporate the probabilistic classification (soft decision making) of the entries within the transformation by updating eq. (4.3) as

$$\epsilon_q = \sum_{t=1}^{N} \left\| \mathbf{y}_t - \mathbf{H}_{\mathbf{q}}^T\widetilde{\mathbf{X}}_t \right\|^2 p(C_q|\mathbf{x}_t) \tag{4.5}$$

where $p(C_q|\mathbf{x}_t)$ is the posterior probability of the qth class C_q, given the input vector $\mathbf{x}_t$, and $\mathbf{H}_{\mathbf{q}}$ is the $(2k+1)(n+1) \times n$ transformation matrix for C_q, which is computed as in eq. (4.4), with $\mathbf{R}_{\mathbf{xx}} = \Sigma_{t=k}^{N-k}\widetilde{\mathbf{X}}_t\widetilde{\mathbf{X}}_t^T p(C_q|\mathbf{x}_t)$ and $\mathbf{r}_{\mathbf{yx}} = \Sigma_{t=k}^{N-k}\mathbf{y}_t\widetilde{\mathbf{X}}_t^T p(C_q|\mathbf{x}_t)$. The transformed feature vector is then calculated by summing the outputs of all the classes as [29]

$$\hat{\mathbf{y}}_t = \Phi(\mathbf{x}_t) = \sum_{q=0}^{Q-1} p(C_q|\mathbf{x}_t)\,\mathbf{H}_{\mathbf{q}}^T\widetilde{\mathbf{X}}_t, \tag{4.6}$$

where Q denotes the total codebook size.

4.3.2 Implementation

Based on the mathematical formulation described in the previous section, the block diagram of the proposed probabilistic transformation system for improving perceptual quality of the PMIC speech, by transforming the signal to the CTM acoustic space, is depicted in Figure 4.4.

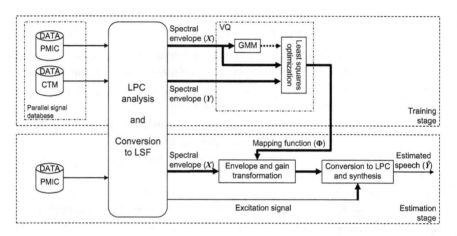

Figure 4.4: Block diagram of the proposed scheme for estimating CTM speech from PMIC speech.

According to the linear source-filter model of speech production, the speech signal can be represented as

$$S(\omega) = E(\omega) \cdot G \cdot V(\omega), \tag{4.7}$$

where $S(\omega)$ is the speech spectrum, $E(\omega)$ is the source excitation, $V(\omega)$ is the vocal tract filter (combined with the lip radiation) and G is a gain factor. In this study, the PMIC speech, represented in the frequency domain as $S^P(\omega)$, is transformed to the CTM speech in the frequency domain as $S^C(\omega)$, by mapping the vocal tract filter $V(\omega)$ and the gain factor G from the PMIC to CTM. The source excitation from the PMIC signal is used to reconstruct the estimated CTM speech. Here, $V(\omega)$ is modeled by AR filter coefficients represented in terms of LSFs [30].

LSFs are often adopted as the feature for artificial bandwidth extension and voice conversion tasks [36, 37], for the following reasons: (i) errors in computing each LSF component do not propagate to other components, (ii) they have good linear interpolation characteristics and are stable [38] and (iii) represented as position and difference coefficients, LSFs are closely related to formant frequency locations and bandwidths [39] (see solid and dashed vertical lines in Figure 4.3). These qualities make LSFs the ideal feature for the current task involving interpolation and enhancement of the degraded frequency components of PMIC speech.

In the training stage (Figure 4.4 top), the LSF and gain feature set is first extracted after LP analysis of the stereo database of PMIC and CTM speech signals. Next, the CTM feature space is vector quantized into Q regions using the binary split LBG algorithm [40]. The distortion measure employed for the VQ is the mean square spectral error of the corresponding LP spectra:

$$D(\mathbf{y}_i^{LSF}, \mathbf{y}_j^{LSF}) = \int\limits_0^{f_0} \left| \mathbf{Y}_i^{LP}(f) - \mathbf{Y}_j^{LP}(f) \right|^2 df, \tag{4.8}$$

where i and j are frame indices, and f_0 is the half sampling frequency.

Subsequently, the posterior probability of each VQ region $p(C_q|\mathbf{x}_t)$ is calculated using Bayes' theorem as follows:

$$p(C_q|\mathbf{x}_t) = \frac{p(C_q)p(\mathbf{x}_t|C_q)}{\sum_{q=0}^{Q-1} p(C_q)p(\mathbf{x}_t|C_q)}, \tag{4.9}$$

where $p(\mathbf{x}_t|C_q)$, which represents the likelihood of the conditioning feature vector $\mathbf{x}_t$ given the VQ region C_q, is modeled as a mixture of Q Gaussian distributions with diagonal covariance matrices, and $p(C_q)$ is the prior probability of the region estimated based on relative frequency from the training data. Here, the posterior probabilities for both the LSF and gain parameters are calculated using only the LSF feature vectors, that is, the Gaussians are conditioned on the LSF feature vectors.

Note that for estimation of the posterior probabilities one could alternatively use the concatenated feature vector, $\widetilde{\mathbf{X}}_t$, (as opposed to $\mathbf{x}_t$) or its dimensionality reduced version after principal component analysis or linear discriminant analysis.

Next, a unique mapping function $\mathbf{H}_q$ is calculated for each VQ region using a neighborhood of $2k+1$ feature vectors, $\widetilde{\mathbf{X}}_t$, through eq. (4.4). Here, two independent mapping functions are computed for mapping the LSF feature vectors and the AR filter gain.

During the estimation phase (see Figure 4.4 bottom), to estimate the CTM feature vectors via eq. (4.6), the LSF feature vectors and gain factor from the PMIC data are used in conjunction with the mapping functions, $\mathbf{H}_q$, and the posterior probabilities, $p(C_q|\mathbf{x}_t)$, which are calculated during the training phase. Given that the LSF parameters take on values between $(0, \pi)$, a linear interpolation of neighboring frames is used to ensure that no LSF component has values outside of this range, thereby alleviating the abnormalities arising from the transformation. Finally, the estimated LSFs are converted back to AR filter coefficients and the estimated CTM speech is resynthesized by exciting the AR filter with the corresponding source signal extracted from the input PMIC data.

Adopting the probabilistic formulation with discrete levels not only offers all the advantages of the traditional VQ approach, but also compensates for the inherent hard decision effects (i.e., one-to-one mapping) by turning the solution into a soft-decision process.

Before concluding this section, it is worth noting that in this chapter, the problem of transforming the source excitation is not considered since it is reliably captured by the PMIC (also see Section 2). In the work reported in Quatieri et al. [22] and Brady et al. [41], it has been shown that natural sounding speech can be produced by using the excitation information from PMIC for the multi-sensor MELPe coder, especially in high-noise environments. In addition, the focus of this chapter is to obtain the best-quality speech from PMIC sensor pickup. As discussed previously, the nonacoustic (PMIC) sensor performance is far superior to a CTM in harsh high-noise conditions versus the difference between PMIC and CTM in clean conditions. Given the obvious benefits of utilizing the nonacoustic PMIC over the CTM under harsh-noise conditions, we choose to concentrate here on the more challenging task (in terms of *preference*) of analysis and quality conversion of the PMIC speech under clean conditions where its quality and intelligibility are both objectively and subjectively scored significantly lower compared to the CTM speech (e.g., if we were to enhance the PMIC output versus CTM in high noise conditions, it would be a trivial task, since even a marginal improvement in the PMIC output would be far superior to the noisy CTM speech). Note that the transformation task itself is much more challenging under noisy conditions due to leakage of acoustic noise through body vibrations and changes in speech production based on feedback from environment (e.g., Lombard effect). This is, however, not our focus in this chapter.

To demonstrate our point even more clearly, spectrograms and waveforms are shown in Figure 4.5 for the same utterance as in Figure 4.2 recorded in broadband speech-shaped noise at a level of 95 dBA-SPL. It is clearly seen from the figure that the spectrotemporal structure of the acoustic CTM signal (Figure 4.5(b)) is severely corrupted due to the background noise, while the nonacoustic PMIC signal (Figure 4.5(a)) is virtually intact. As a result, by

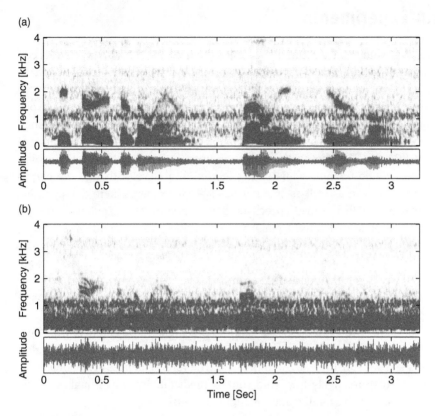

Figure 4.5: Spectrotemporal comparison between (a) PMIC and (b) CTM signals for the utterance "be sure to set the lamp firmly in the hole," recorded in speech-shaped noise at 95 dBA SPL.

demonstrating quality improvement with the proposed transformed PMIC to PMIC solution in the "noise-free" domain, if this is successful, the solution will be far superior for the noisy case. Since one of the performance criteria is based on human listener evaluations, comparing the enhanced PMIC signal captured in noisy conditions and comparing this to the noisy CTM signal is a much easier perceptual task than enhancing the PMIC signal captured in clean conditions and comparing to the clean CTM version (i.e., any processing artifacts arising from the transformation would be much more noticeable if the enhanced PMIC is compared to the clean CTM versus noisy CTM versions).

4.4 Experiments

Objective evaluations are carried out using the UTScope speech under stress corpus [42]. The UTScope database consists of 86 speakers (53 females and 33 males). The corpus consists of data recordings from a range of domains; the three of interest here are the neutral task (relaxed sitting), cognitive task (involving the use of a racing-car simulator) and physical task (involving exercise stairstepping). For this study, we focus on the neutral task with stereo recordings of the PMIC and CTM. The neutral task recordings were collected with the subject repeating audio prompts while sitting relaxed in a chair. Although the corpus comprises of read as well as spontaneous speech segments, here the algorithm evaluation is with the read speech portion consisting of 35 sentence prompts from three male and three female subjects (subject IDs – mml1, mss1, mwl1, fac1, fah1 and fth1). These subjects are all native speakers of American English. The audio prompts are similar to those used in the TIMIT database, and all approximately 2–5 s in duration. Due to the lack of sufficient training data for speaker-independent modeling (35 sentences per speaker, which are 2–5 s long), we perform our experiments using speaker-dependent transformations.

For subjective evaluations, a total of 60 sentences (6 sentence lists) from the IEEE database [43] were recorded simultaneously from an American adult male speaker using the PMIC and the CTM. The IEEE sentences are phonetically balanced with relatively low word-context predictability, which makes it well suited for subjective quality and intelligibility tests.

Data processing is set up as follows: First, LSF coefficients are obtained from a 12th-order LPC analysis of speech signals (assuming sampling frequency of 8 kHz). The analysis and synthesis are performed using the overlap-and-add technique with frames of 25 ms duration at a rate of 100 frames per second.

For the experiments, spectral feature vectors (LSF) and AR filter gain are mapped separately. The AR filter gain estimation is crucial in mapping PMIC to CTM as it plays an important role in shaping the overall synthesized speech waveform. Reliable estimation of this parameter can amplify those parts of the PMIC speech attenuated due to the nonlinear filtering effect of the contact interface. The length of $\widetilde{\mathbf{X}}_t$ is fixed at five feature frames ($k = 2$).

To evaluate the probabilistic mapping function, a *leave-one-out* (LOO) cross-validation scheme is adopted such that out of the 35 sentence prompts from the UTScope, 34 sentences are used to construct the mapping function, while the remaining sentence is employed for evaluation. In the case of IEEE sentences, from a total of six sentence lists, sentences from five lists are used for training and the remaining sentence set is employed for evaluation. The final performance measure of the transformation algorithm is obtained by averaging results across all evaluations. The LOO strategy is adopted to ensure that the maximum amount of training data is used and that the results are not biased due to data partitioning.

The quality of the estimated speech signals is objectively evaluated using five distinct metrics. The perceptual evaluation of speech quality (PESQ) measure defines the difference between the loudness spectra averaged over time and frequency of a time-aligned target and enhanced signal. Furthermore, as the two signals are equalized to a standard listening level, the PESQ is highly correlated with the mean opinion score (MOS). The frequency-weighted segmental SNR (fwSNRseg) is another measure highly correlated with the subjective listening tests [44]. By computing a frequency-weighted SNR, the problem of obtaining infinity via a segmental SNR computation of a silence segment is avoided. It has been shown that both the PESQ and fwSNRseg measures perform modestly well in terms of predicting both quality and intelligibility [45]. The log-likelihood ratio (LLR) and Itakura–Saito distance (ISD) measures are based on representing the dissimilarity between the AR models of the target and the enhanced speech signals. The LLR represents the ratio of the prediction residual energies of the two signals, while the ISD emphasizes the differences in the gains of the AR models. The weighted spectral slope (WSS) distance measure [46] is designed to magnify the differences in the spectral peak locations between the target and the enhanced signals. A study on perceived distance between vowels [47] indicated that differences in formant frequencies is the most differentiating factor as compared to other spectral factors like filtering, formant bandwidths and spectral tilt. Thus, an algorithm that adjusts the spectral peaks closer to the target speech will score higher on the WSS measure (for more details on these metrics, see Loizou [11], Quankenbush et al. [48]). The LLR, ISD and WSS have been evaluated for 330 different types of distortions in [48]. Since each metric is formulated based on a particular speech feature, it is clear that no single metric is optimal across all conditions [48]. Therefore, to have a reliable performance evaluation, the results are represented in terms of the above-mentioned metrics.

To subjectively evaluate overall quality and intelligibility of the estimated speech, a total of 14 normal-hearing American listeners were recruited, paid and trained for the listening experiments. The experiment was performed in a sound-proof room. Two lists of sentences (20 sentences) were used per signal type (i.e., the original PMIC, processed PMIC and CTM). Stimuli were played to the listeners through open-air Sennheiser HD580 headphone at their desired volume levels. Subjects were asked to judge overall speech quality as well as their own listening effort. They were also asked to transcribe what they heard. The appearance of sentences from different signal types was randomized to ensure that the results are not biased due to listeners' predictions.

4.5 Results and discussion

In this section, performance of the proposed mapping scheme is evaluated based on four domains: (i) spectral and temporal waveform analysis, (ii) LP spectra and LSFs of the phonemes, (iii) five distinct objective measures (PESQ, fwSNRseg, LLR, ISD and WSS) and (iv) subjective quality and intelligibility (see [49], Quackenbush et al., 2001, [48], for details on these measures).

Spectrograms for the utterance "be sure to set the lamp firmly in the hole" of the estimated and CTM signals are illustrated in Figure 4.6. As described in Section 2, the PMIC signal is a low-passed version of the CTM signal, and the spectrogram in Figure 4.2(a) confirms the weak presence of the frequency components in 1–3 kHz range. After applying the probabilistic mapping to transform the PMIC speech to that of the CTM, it can be observed from Figure 4.6(a)

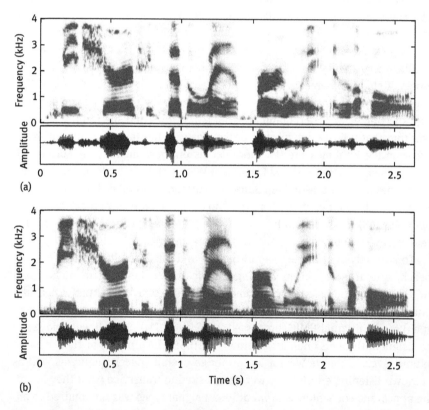

Figure 4.6: Spectrotemporal comparison between (a) estimated and (b) CTM signals for the utterance "be sure to set the lamp firmly in the hole.".

that the complete frequency band of 1 kHz to 3 kHz and beyond is effectively reconstructed, and the spectral richness in the estimated signal is clearly depicted.

More specifically, the sound /sh/ between 250 and 400 ms, which is almost absent in Figure 4.2(a), is fairly reconstructed, and the frequency content around 2 kHz and above is clearly visible in the estimated CTM signal (Figure 4.6(a)). Moreover, the formant transitions in the 1000–1300 ms range are strengthened with the help of the proposed transformation. Thus, based on spectrogram and temporal analysis, it is obvious that the proposed algorithm has the potential to drastically improve the spectral content and richness in the estimated signal, bringing it close to the expected CTM speech signal.

Next, algorithm performance across the phonemes presented in Section 2 (Figure 4.3) is illustrated in Figure 4.7. The LP spectra and the corresponding LSFs (represented in terms of solid and dashed vertical lines) of the front vowel /i/, back vowel /u/, nasal consonant /n/ and fricative consonant /sh/ are shown. In general, as can been seen from these two figures (Figures 4.3 and 4.7), the PMIC LSFs are mapped much closer to that of the CTM via the proposed algorithm. Since the LSFs are closely related to the formant locations and their bandwidths, the formants of the estimated speech are also closer to that of the CTM speech.

Figure 4.7(a) represents the LP spectra for the front vowel /i/. The third formant that was incorrectly shifted close to the fourth formant (see Figure 4.3(a)) is now moved back to 2.7 kHz as seen from Figure 4.7(a). The level of the LP spectra is also adjusted in the vicinity of the first formant. Part (b) of the two figures represents the back vowel /u/. The formant F3, which is completely missing in the PMIC spectrum, is reconstructed, and the formant F4, which is relatively boosted, is attenuated to the proper CTM formant level. Also, the spectral slope is adjusted so that F4 is now lower than F3 and helps in matching the CTM formant contour.

The LP spectra for the nasal consonant /n/ are shown in Figure 4.7(c). The undesired broadening of the first formant (likely due to a shift of the second formant) as well as the presence of the fourth formant, as observed in Figure 4.3(c), are both suppressed by the proposed mapping scheme. In addition, bandwidth sharpening at F1 and F3 as well as reconstruction of F2 are depicted in Figure 4.7(c). Thus, the estimated spectrum for the sound /n/ is much closer to the CTM spectrum, although the complete recovery of the LP spectra is still missing. The algorithm performance for the fricative consonant /sh/ is illustrated in Figure 4.7(d). Although F2 is still not reconstructed faithfully, F3 and F4 are transformed closely to the CTM LP spectrum. Therefore, based on the individual phoneme analysis via the LP spectra and LSFs, the proposed algorithm provides considerable improvement.

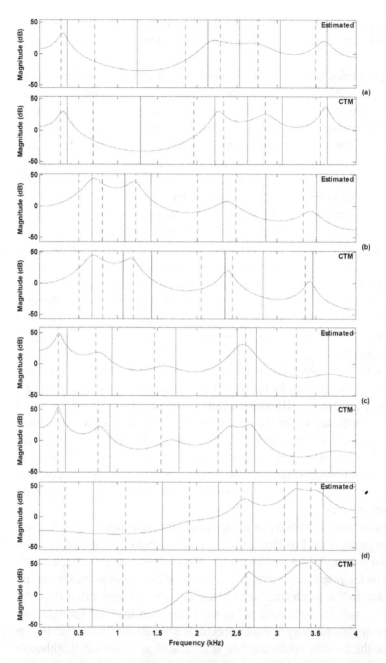

Figure 4.7: The averaged LP spectra and corresponding LSFs for the (a) front vowel/i/, (b) back vowel/u/, (c) nasal consonant /n/ and (d) fricative consonant /sh/.

To investigate the impact of an increase in the VQ codebook size on mapping performance, the codebook size is varied from 2 to 8 bits and the results are presented in terms of the objective measures. Table 4.1 lists the average performance based on the PESQ and ISD measures obtained using LOO cross-validation scheme. The PESQ is valued between –1 and 4.5, with –1 indicating poor quality and 4.5 the best quality. The ISD measures the perceptual difference between the source and target spectra [48]. Smaller values of ISD indicate greater similarity in speech spectra. From the table, for most cases, the two measures indicate improvement as compared to the baseline (i.e., the metric computed for the original PMIC and CTM) for all the speakers with an increase in codebook size until 6 bits, after which performance tappers off or remains constant. Informal listening tests also confirm improvement in perceptual speech quality for codebook sizes of 2 to 6 bits, while some glitches are observed in the resynthesized speech for 7-bit and 8-bit codebooks. These glitches are because of the scarcity of speech data (34 sentences each between 2 and 5 s is not sufficient to train a codebook of that size) in some VQ regions, making the autocorrelation matrix $\mathbf{R}_{xx}$ ill-conditioned, causing the AR filter to become unstable.

Table 4.1: Average objective measures (PESQ and ISD) for three female and three male speakers.

Speaker ID		VQ size (bits)							
		Baseline	2	3	4	5	6	7	8
fac1	PESQ	2.13	2.07	2.11	2.22	2.41	**2.55**	2.46	2.46
	ISD	2.03	1.00	0.94	0.92	0.89	0.87	**0.84**	0.87
fah1	PESQ	1.76	2.18	2.19	2.34	2.75	**2.98**	2.80	2.66
	ISD	5.99	1.12	1.07	1.03	1.02	1.00	**0.99**	1.02
fth1	PESQ	2.03	2.26	2.37	2.60	2.70	**2.91**	2.65	2.52
	ISD	2.79	0.74	0.66	0.61	0.58	0.56	**0.36**	0.55
mml1	PESQ	1.98	2.21	2.26	2.55	**2.66**	2.48	2.48	2.44
	ISD	2.23	0.70	0.73	0.69	0.66	0.66	0.64	**0.28**
mss1	PESQ	1.97	2.30	2.39	2.38	**2.95**	2.88	2.80	2.64
	ISD	2.84	0.74	0.68	0.66	0.63	0.61	0.57	**0.49**
mwl1	PESQ	2.05	2.17	2.31	2.55	2.75	**2.88**	2.81	2.81
	ISD	2.42	0.65	0.61	0.58	0.56	0.54	**0.38**	0.52

Figure 4.8 shows the ISD measure for CTM and PMIC spectra (dashed lines) and CTM and the estimated spectra (solid lines) over time for a sample speech utterance. The distances between the estimated and CTM spectra are very small when compared to the source-target spectra. This indicates that the estimated spectra are close to that of the CTM. Thus, the proposed probabilistic mapping is also capable of capturing the spectral correlation between PMIC and CTM stereo data.

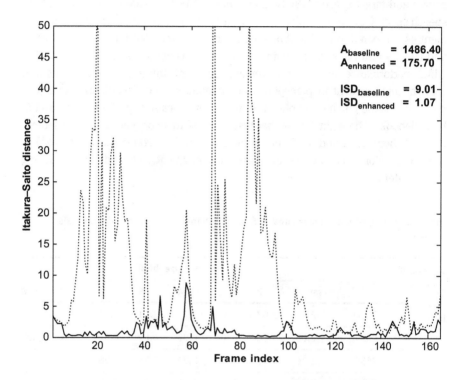

Figure 4.8: Itakura-Saito distance (ISD) between CTM and PMIC spectra (dashed lines) and CTM and estimated spectra (solid lines) for a sample speech utterance. Areas under the baseline and enhanced ISD contours are 1486.4 and 175.7, respectively. Average baseline and enhanced ISD measures are 9.01 and 1.07, respectively.

To observe the effect of an increase in VQ codebook size on the performance of the transformation algorithm more clearly, the average LLR measure is depicted in Figure 4.9 for different speakers and codebook sizes. Similar to the ISD measure, smaller values of this metric imply a greater match between the two spectra. It can be seen from the figure that the values for this measure almost follow the same trend as the ISD measure. In addition, it is interesting to note that even with a small codebook size (2 bits) the system can achieve acceptable performance,

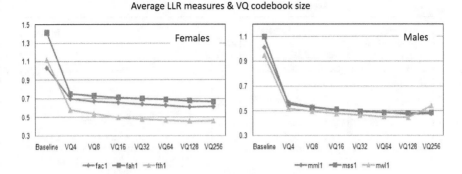

Figure 4.9: Average LLR measure for the three female (left) and three male (right) speakers for different VQ codebook sizes.

although further increasing the codebook size provides better results. For this measure, the mwl1 speaker is influenced by the data scarcity effect more than the others (performance degrades for the 8-bit codebook).

Figure 4.10 demonstrates average fwSNRseg measure for the three female and three male speakers for different codebook sizes. The measure signifies the critical band-based segmental SNR performance. As compared to the LLR measure (Figure 4.9), fwSNRseg shows relatively steeper performance improvement with an increase in the codebook size. It is worth noting that the similar trend is observed with the PESQ measure. The performance represented in terms of fwSNRseg and LLR measures indicates that the fth1 speaker benefits more with

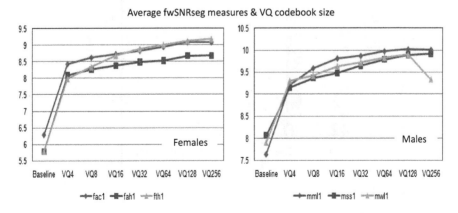

Figure 4.10: Average fwSNRseg measure for the three female (left) and three male (right) speakers for different VQ codebook sizes.

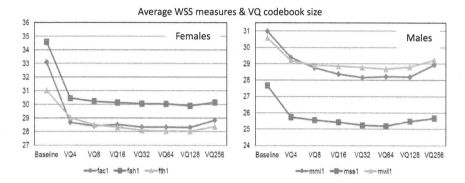

Figure 4.11: Average WSS measure for the three female (left) and three male (right) speakers for different VQ codebook sizes.

an increase in the codebook size as compared to the other speakers in the group.

The WSS [48] measure mimics the importance of the overall spectral slope. Smaller values of this measure represent overall better algorithm performance. As depicted in Figure 4.11, the measure variability for different codebook sizes is more evident when compared to the other performance metrics. The measure values reduce by increasing the codebook size until 7 bits; the smallest values (local minima at a particular codebook size) agree with that obtained for the PESQ.

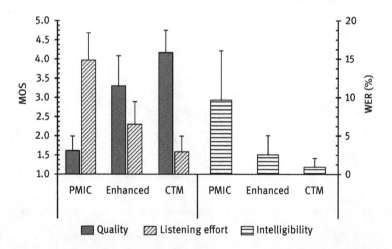

Figure 4.12: Subjective assessment of overall quality and intelligibility of the transformed signals in terms of MOS for speech quality and listening effort as well as mean WER.

Figure 4.12 demonstrates the results of the listening experiment in terms of perceived quality and listening effort (left half) as well as WER scores (right half), averaged across the subjects. As can be seen from the left half of the figure (primary vertical axis), there is a dramatic elevation in overall quality of the enhanced PMIC signals compared to the original recordings. The amount of listening effort is also relatively decreased for the processed signals. As expected, the human-based WER scores (secondary vertical axis) indicate a decline in intelligibility of the PMIC recordings compared to that of the CTM speech. It is observed that the proposed approach has successfully improved the intelligibility of the original PMIC signals. In all the cases, improvements are statistically significant (paired t-test: $p < 0.001$).

4.6 Conclusion

This study has considered a comprehensive acoustic analysis of PMIC captured speech, and explored a signal processing strategy for improving speech quality and intelligibility of non-acoustic signal capture for human-to-human communications. A continuous probabilistic transformation technique was proposed to enhance overall quality and intelligibility of signals measured from the nonacoustic PMIC. The algorithm performance was evaluated by careful analysis of spectral structure across the utterances, as well as individual phonemes. The proposed scheme exhibited significant improvement over a range of objective measures that predict perceptual quality and intelligibility of speech signals faithfully – PESQ, ISD, LLR, fwSNRseg and WSS. All tested measures indicated that the proposed transformation scheme proved to be useful in reconstructing the nonlinear-frequency-mapped PMIC speech spectra. Moreover, subjective listening experiments were also carried out to assess (i) overall quality, (ii) listening effort and (iii) intelligibility of the transformed PMIC signals, and the outcomes indicated the probabilistic transformation method is quite effective in removing the metallic and muffled nature prevalent in traditional PMIC signals, as well as increasing intelligibility. The results obtained in this study, which considered a standard spectral transformation approach for quality conversion of the PMIC speech, motivate further research on development of more sophisticated transformation techniques for the task. It also suggests alternatives to improving human-to-human speech communications in subjects who possess vocal fold pathologies, restricted vocal fold mechanics or excitation limitations due to reduced forced vital capacity of the lungs. While further research is necessary to transition such an approach to real-time applications, the algorithmic approach has been shown to be promising.

References

[1] Assmann, P. F., and Summerfield, A. Q. The perception of speech under adverse conditions, Greenberg, S., Ainsworth, W. A., Popper, A. N., Fay, R. R, Eds., Speech Processing in the Auditory System, Springer-Verlag, New York, Ch. 5, 231–308, 2004.

[2] Hansen, J. H. L., and Pellom, B. An effective quality evaluation protocol for speech enhancement algorithms, Proc. ICSLP'98, Sydney, Australia, 2819–2822, 1998.

[3] van Rooji, J., and Plomp, R. Auditive and cognitive factors in speech perception by elderly listeners. III. Additional data and final discussion. J. Acoust. Soc. Am, 91(2), 1028–1033, 1992.

[4] Picheny, M. A., Durlach, N. I., and Braida, L. D. Speaking clearly for the hard of hearing I: Intelligibility differences between clear and conversational speech. J. Speech Hear. Res., 28(1), 96–103, 1985.

[5] Junqua, J.-C. The Lombard reflex and its role on human listeners and automatic speech recognizers. J. Acoust. Soc. Am., 93(1), 510–524, 1993.

[6] Gavidia-Ceballos, L., and Hansen, J.H.L. "Direct speech feature estimation using an iterative EM algorithm for vocal cancer detection." IEEE Transactions on Biomedical Engineering, 43(4), 373–383, April 1996.

[7] Hansen, J.H.L., Gavidia-Ceballos, L., and Kaiser, J.F. "A nonlinear operator-based speech feature analysis method with application to vocal fold pathology assessment." IEEE Transactions on Biomedical Engineering, 45(3), 300–313, March 1998.

[8] Godino-Llorente, J. I., and Gómez-Vilda, P. "Automatic detection of voice impairments by means of short-term cepstral parameters and neural network based detectors." IEEE Transactions on Biomedical Engineering, 51(2), 380–384, Feb 2004.

[9] Cairns, D., Hansen, J.H.L., and Riski, J. 1996. "A noninvasive technique for detecting hypernasal speech using a nonlinear operator." IEEE Transactions on Biomedical Engineering, 43, 35–45, Jan 1996.

[10] Godino-Llorente, J. I., Gómez-Vilda, P., and Blanco-Velasco, M. "Dimensionality reduction of a Pathological Voice Quality Assessment System Based on Gaussian Mixture Models and short-term cepstral parameters." IEEE Transactions on Biomedical Engineering, 53(10), 1943–1953, Oct 2006.

[11] Loizou, P. C. Speech Enhancement: Theory and Practice, CRC Press, Boca Raton, FL, 2007.

[12] Hansen, J.H.L. "Speech Enhancement," invited contributor, Encyclopedia of Electrical and Electronics Engineering, John Wiley & Sons, Vol. 20, 159–175, 1999.

[13] Bass, J. D., Scanlon, M. V., and Mills, T. K. Getting two birds with one phone: An acoustic sensor for both speech recognition and medical monitoring. J. Acoust. Soc. Am., 106(4), 2180, 1999.

[14] Ingalls, R. Throat microphone. J. Acoust. Soc. Am., 81(3), 809, 1987.

[15] Nakajima, Y., Kashioka, H., Campbell, N., and Shikano, K. Non-audible murmur (NAM) recognition. IEICE Trans. Inf. Syst, E89–D(1), 1–4, 2006.

[16] Toda, T., Nakamura, K., Sekimoto, H., and Shikano, K., 2009. Voice conversion for various types of body transmitted speech. In: Proc. IEEE ICASSP'09. Taipei, Taiwan, pp. 3601–3604.

[17] Toda, T., Nakagiri, M., and Shikano, K. Statistical voice conversion techniques for body-conducted unvoiced speech enhancement. IEEE Trans. Audio Speech Lang. Process, 20(9), 2505–2517, 2012.

[18] Yanagisawa, T., and Furihata, K. Pickup of speech signal by utilization of vibration transducer under high ambient noise. J. Acoust. Soc. Jpn., 31(3), 213–220, 1975.

[19] Burnett, G. C., Holzrichter, J. F., Ng, L. C., and Gable, T. J. The use of glottal electromagnetic micropower sensors (gems) in determining a voiced excitation function. J. Acoust. Soc. Am., 106(4), 2183–2184, 1999.

[20] Brown, D. R., Keenaghan, K., and Desimini, S. Measuring glottal activity during voiced speech using a tuned electromagnetic resonating collar sensor. Meas. Sci. Technol., 16(11), 2381–2390, 2005.

[21] Subramanya, A., Zhang, Z., Liu, Z., and Acero, A. Multisensory processing for speech enhancement and magnitude-normalized spectra for speech modeling. Speech Commun., 50(3), 228–243, 2008.

[22] Quatieri, T. F., Brady, K., Messing, D., Campbell, J. P., Campbell, W. M., Brandstein, M. S., Weinstein, C. J., Tardelli, J. D., and Gatewood, P. D. Exploiting nonacoustic sensors for speech encoding. IEEE Trans. Audio Speech Lang. Process, 14 (2), 533–544, 2006.

[23] Graciarena, M., Franco, H., Sonmez, K., and Bratt, H. Combining standard and throat microphones for robust speech recognition. IEEE Signal Process. Lett., 10(3), 72–74, 2003.

[24] Singh, A., Sangwan, A., and Hansen, J. H. L., 2012. Improved parcel sorting by combining automatic speech and character recognition. In: Proceedings of IEEE International Conference on Emerging Signal Processing Applications, (ESPA '12). Las Vegas, NV, pp. 52–55.

[25] Patil, S. A., and Hansen, J. H. L. The physiological microphone (PMIC): A competitive alternative for speaker assessment in stress detection and speaker verification. Speech Commun., 52(4), 327–340, 2010.

[26] Tran, V.-A., Bailly, G., Loevenbruck, H., and Toda, T. Improvement to a NAM-captured whisper-to-speech system. Speech Commun., 52(4), 314–326, 2010.

[27] Shahina, A., and Yegnanarayana, B. Mapping speech spectra from throat microphone to close-speaking microphone: A neural network approach. EURASIP J. Audio Speech Music Process, 2007(2), 1–10, 2007.

[28] Zhang, Z., Liu, Z., Sinclair, M., Acero, A., Deng, L., Droppo, J., Huang, X., and Zheng, Y., 2004. Multi-sensory microphones for robust speech detection, enhancement and recognition. In: Proc. IEEE ICASSP'04. Vol. 3. Montreal, QC, Canada, pp. 781–784.

[29] Neumeyer, L., and Weintraub, M., 1994. Probabilistic optimum filtering for robust speech recognition. In: Proc. IEEE ICASSP'94. Vol. 1. Adelaide, Australia, pp. 417–420.

[30] Itakura, F. Line spectral representation of linear prediction coefficients of speech signals. J. Acoust. Soc. Am., 57(1), S35, 1975.

[31] Abe, M., Nakamura, S., Shikano, K., and Kuwabara, H. Voice conversion through vector quantization. J. Acoust. Soc. Jpn, E-11(2), 71–76, 1990.

[32] Scanlon, M. Acoustic sensor for voice with embedded physiology, Military Sensing Symp, North Charleston, SC, USA, 1999.

[33] French, N., and Steinberg, J. Factors governing the intelligibility of speech sounds. J. Acoust. Soc. Am., 19(1), 90–119, 1947.

[34] Nakatani, L., and McDermott, B. Effect of pitch and formant manipulation on speech quality. J. Acoust. Soc. Am., 50(1A), 145, 1971.

[35] Patterson, R., and Moore, B. Auditory filters and excitation patterns as representations of frequency resolution, Moore, B. C. J., Ed., Frequency Selectivity in Hearing, Academic Press, London, 1986.

[36] Enbom, N., and Kleijn, W. B., 1999. Bandwidth expansion of speech based on vector quantization of the mel frequency cepstral coefficients. In: Proc. IEEE Workshop on Speech Coding. Porvoo, Finland, pp. 171–173.

[37] Pellom, B., and Hansen, J. H. L., 1997. Spectral normalization employing hidden Markov modeling of line spectrum pair frequencies. In: Proc. IEEE ICASSP'97. Vol. 2. Munich, Germany, pp. 943–946.

[38] Paliwal, K., 1995. Interpolation properties of linear prediction parametric representations. In: Proc. EuroSpeech'95. Vol. 2. Madrid, Spain, pp. 1029–1032.

[39] Crosmer, J., 1985. Very low bit rate speech coding using the line spectrum pair transformation of the LPC coefficients. Ph.D. thesis, Georgia Institute of Technology, Atlanta, GA.

[40] Linde, Y., Buzo, A., and Gray, R. An algorithm for vector quantizer design. IEEE Trans. Commun., 28(1), 84–95, 1980.

[41] Brady, K., Quatieri, T., Campbell, J., Campbell, W., Brandstein, M., and Weinstein, C., 2004. Multisensor MELPe using parameter substitution. In: Proc. IEEE ICASSP'04. Vol. 1. pp. 477–480.

[42] Ikeno, A., Varadarajan, V., Patil, S., and Hansen, J.H.L. UT-Scope: Speech under Lombard effect and cognitive stress, Proc. IEEE Aerosp. Conf, Vol. 7, Big Sky, MT, USA, 1–7, 2007.

[43] IEEE, 1969 IEEE recommended practice for speech quality measurements. IEEE Trans. Audio Electroacoust, AU-17(3), 225–246.

[44] Wang, S., Sekey, A., and Gersho, A. An objective measure for predicting subjective quality of speech coders. IEEE J. Sel. Areas Commun., 10(5), 819–829, 1992.

[45] Ma, J., Hu, Y., and Loizou, P. C. Objective measures for predicting speech intelligibility in noisy conditions based on new band-importance functions. J. Acoust. Soc. Am., 125(5), 3387–3405, 2009.

[46] Klatt, D., 1982. Prediction of perceived phonetic distance from critical-band spectra: A first step. In: Proc. IEEE ICASSP'82. Vol. 7. Paris, France, pp. 1278–1281.

[47] Wang, M., and Bilger, R. Consonant confusions in noise: A study of perceptual features. J. Acoust. Soc. Am., 54(5), 1248–1266, 1973.

[48] Quackenbush, S., Barnwell, T., and Clements, M. Objective Measures of Speech Quality, Prentice-Hall, New Jersey, 1988.

[49] Rix, A.W., Beeremds, J.G., Hollier, M.P., and Hekstra, A.P., 2001. "Perceptual evaluation of speech quality (PESQ)-a new method for speech quality assessment of telephone networks and codecs," IEEE ICASSP-2001, pp. 749–752.

[50] Sadjadi, S. O., Patil, S. A., and Hansen, J. H. L., 2010. Quality conversion of non-acoustic signals for facilitating human-to-human speech communication under harsh acoustic conditions. In: Proc. INTERSPEECH-2010. Makuhari, Japan, pp. 1624–1627.

Nirmesh J. Shah and Hemant A. Patil

5 Non-audible murmur to audible speech conversion

Abstract: In the absence of vocal fold vibrations, movement of articulators which produced the respiratory sound can be captured by the soft tissue of the head using the nonaudible murmur (NAM) microphone. NAM is one of the silent speech interface techniques, which can be used by the patients who are suffering from the vocal fold-related disorders. Though NAM microphone is able to capture very small NAM produced by the patients, it suffers from the degradation of quality due to lack of radiation effect at the lips and the lowpass nature of the soft tissue, which attenuate the high-frequency-related information. Hence, it is mostly unintelligible. In this chapter, we propose to use deep learning-based techniques to improve the intelligibility of the NAM signal. In particular, we propose to use deep neural network-based conversion technique with rectifier linear unit as the nonlinear activation function. The proposed system converts this less intelligible NAM signal to the audible speech signal. In particular, we propose to develop a two-stage model where the first stage converts the NAM signal to the whispered speech signal and the second model will convert this whispered speech signal to the normal audible speech. We compared the performance of our proposed system with respect to the state-of-the-art Gaussian mixture model (GMM)-based method. From the objective evaluation, we found that there is 25% of relative improvement compared to the GMM-based NAM-to-whisper (NAM2WHSP) system. In addition, our second-stage speaker-independent whisper-to-speech conversion system further helps NAM-to-speech conversion system to extract linguistic message present in the NAM signal. In particular, from the subjective test, we found 4.15% of the absolute decrease in word error rate compared to the NAM2WHSP system.

Keywords: Nonaudible murmur, whisper, deep neural network, NAM-to-whisper, whisper-to-speech, voice conversion

Nirmesh J. Shah, Hemant A. Patil, Dhirubhai Ambani Institute of Information and Communication Technology (DA-IICT), India

https://doi.org/10.1515/9781501501265-006

5.1 Introduction

Patients suffering from the vocal fold-related disorders, such as vocal fold paresis (i.e., partial paralysis) or paralysis [1, 2] and vocal nodule [3, 4], may not be able to produce normal audible speech due to the partial or complete absence of vocal fold vibrations. These result into glottis being open during phonation. Hence, noisy excitation source will be generated, which will result in a breathy or whispered voice. Loosing the natural way of producing speech will affect tremendously the patients' life since speech is one of the most usual ways of communication among humans. These low-power articulator sounds without or partial vocal fold vibrations can be considered as silent speech or murmuring [5, 6]. Capturing this silent speech accurately and converting it into the audible sound is the goal of this chapter. Among the various available silent speech interfaces [5, 7], such as electromagnetic articulography sensor [8], throat microphone [9], electromyography (EMG) [10, 11], surface EMG of the larynx [12, 13], ultrasound and optical imaging of the tongue or lips [14–16] and nonaudible murmur (NAM) microphone [6, 17, 18], we have selected the NAM microphone in this chapter due to its simple usability [19]. In the absence of vocal fold vibrations, the respiratory sound produced by the movement of articulators can be recorded via the soft tissue of the head without any obstructions from the bones. Such a recording is called nonaudible murmur [6]. Despite being the potential device for capturing the silent speech, it suffers from degradation of quality due to lack of radiation effect at the lips and the lowpass nature of the soft tissue, which attenuate the high-frequency-related information. Hence, conversion of NAM speech to audible speech is a very challenging or formidable task [20]. Broadly, this NAM-to-audible speech conversion is divided into two approaches. One uses the speech recognizer and the speech synthesis-based approach [16] and the other uses conversion or mapping techniques [21–23]. The earlier approach requires linguistic information in both synthesis and recognition, which is mostly unavailable and also our goal is to extract the message from that silent speech. Hence, we focused on the latter approach.

Earlier NAM-to-speech conversion (NAM2SPCH) system uses the Gaussian mixture model (GMM)-based conversion techniques [21]. Recently, generative adversarial network-based approach has also become popular for the NAM2WHSP conversion task [24]. In this chapter, we propose to use artificial neural network (ANN) and deep neural network (DNN)-based framework. In particular, we propose to use rectifier linear unit (ReLU) as a nonlinear activation function due to its faster convergence property [25]. Furthermore, due to the unavailability of normal speech utterances corresponding to the NAM utterances from a given speaker, we propose to develop speaker-independent whisper-to-normal speech conversion system. Among available various WHSP2SPCH conversion systems [26–31], we use

GMM, ANN and DNN-based framework in this chapter. Due to our proposed idea, the development of the speaker-independent WHSP2SPCH conversion system, more amount of training data from a number of speakers have been taken, which helps to train DNN-based system very well. We have evaluated our systems using various objective measures and selected the best performing NAM2WHSP and WHSP2SPCH. Hence, we combined two separate systems, namely, NAM2WHSP and WHSP2SPCH to convert NAM-to-audible speech (NAM2SPCH). Here, we have also presented a subjective analysis of the various developed systems.

The organization of the rest of the chapter is as follows. Section 2 briefly describes the medical relevance of the proposed problem. Section 3 presents the details regarding NAM microphone and recording of NAM signal. Section 4 discusses the various state-of-the-art conversion techniques. Section 5 explains the proposed framework for the NAM to audible speech conversion. In addition, it also discusses the proposed architecture of NAM-to-whisper (NAM2WHSP) and whisper-to-speech (WHSP2SPCH) conversion systems. Section 6 explains the experimental setup and discusses the results of the various proposed works presented in this chapter.

5.2 Medical relevance of the problem

patients who are having diseases related to the larynx or voicebox may loose the normal ability to produce speech [32]. In particular, disorders related to the mobility of the vocal folds lead to hoarseness, breathiness, rough voice or whispered speech. A brief description of the vocal fold-related diseases is given as follows:

- **Vocal fold paresis or paralysis** [1, 33]: Partial or total weakness of a vocal nerve is called as vocal fold paresis and vocal fold paralysis, respectively. This happens due to the injury at recurrent laryngeal nerve (RLN), which is primarily the cause for the motor input at the vocal folds. Hence, these diseases are also known as RLN paralysis. In this disorder, one or both vocal folds are immobile due to which one or both vocal folds may not close properly (as shown in Figure 5.1), which leads to glottal insufficiency [34]. Hence, the patient will not be able to speak normally and loudly. The vocal fold paresis or paralysis occurs primarily due to the neck injuries, tumors, and neck and head surgeries. Most commonly, post-thyroid surgery causes this disease [35]. However, sometimes carotid artery surgery or spinal surgery that goes through the front side of the neck will also lead to this disorder [2]. A schematic representation of vocal fold is shown in Figure 5.1 for various vocal disorders.
- **Vocal fold lesions** [36]: These are noncancerous benign growths that include vocal nodules and vocal polyp [3, 37]. The vocal polyp is also known

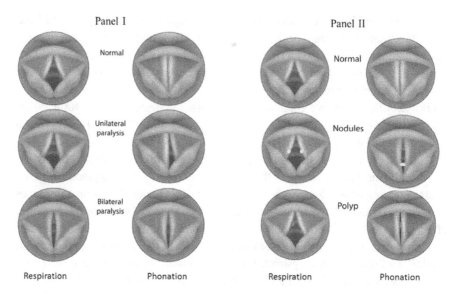

Figure 5.1: Schematic representation of vocal folds during respiration and phonation phase. Panel I: vocal fold paralysis, and panel II: vocal fold lesions. Adapted from [39].

as the Reinke's edema or polypoid degeneration. A vocal polyp is also the benign growth similar to the vocal nodule. However, it is softer than the vocal nodule. This growth on the vocal folds results in an incomplete closure of the vocal folds (as shown in Figure 5.1), which also lead to hoarseness and whispering. The primary source of this disorder is the vocal abuse and vocal misuse [4, 38]. Vocal abuse is any habit that strains the vocal fold, for example, excessive talking, coughing, inhaling irritants, smoking, drinking a lot of alcohol, screaming or yelling. On the other hand, vocal misuse is related to the inappropriate usage of voice, such as speaking too loudly or speaking with lower or higher pitch or fundamental frequency (F_0).

- **Laryngitis:** It is the swelling or the inflammation of the vocal folds. The primary causes of these diseases are the viral or bacterial infections. In addition, it may be caused due to the gastroesophageal reflux (i.e., due to the backup of stomach acid into the throat) and inhaled chemicals.
- **Neurological disorders** [32]: Apart from the earlier mentioned disorders, neurological disorders such as myasthenia gravis, multiple sclerosis, Parkinson's disease and amyotrophic lateral sclerosis (also known as motor neuron disease) also lead to sometimes vocal fold-related diseases.

- **Psychological conditions:** Most commonly hysterical aphonia (brought on by experiencing or seeing a traumatic event) and selective mutism (symptoms of anxiety disorders) also lead to the whispering.

As speech is the most comfortable way of communication among humans, loosing the natural ability of speaking (due to any of the above-mentioned disorders) results in the psychological burden on the patients' lives. In addition, understanding their messages will be difficult for the normal person in day-to-day communication. Hence, we need signal processing and machine learning-based techniques, which convert their breathy voice or whispered voice into normal audible speech and make it more intelligible and this is the goal of this chapter.

5.3 Analysis of NAM signal and whispered speech

5.3.1 NAM signal recording

In the absence of vocal fold vibrations, the respiratory sound produced by the movement of articulators, such as palate, tongue, and lips, can be recorded via the soft tissue of the head without any obstructions from the bones. Such recording is called as nonaudible murmur [6]. NAM is recorded by the special kind of stethoscopic microphone attached below the ear to the surface of the skin as shown in Figure 5.2.

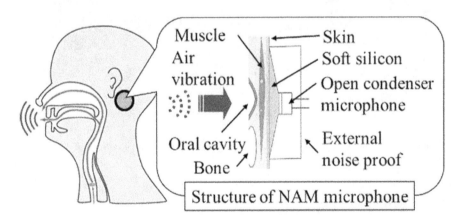

Figure 5.2: Schematic representation of NAM microphone. Adapted from [17, 20].

The earlier tissue-conductive microphone contains an electret condenser microphone, which is covered with urethane elastomer or soft silicone-like soft polymer material for better impedance matching with the soft tissue of the neck [20, 40]. The acoustic vibration of vocal tract system is captured via sensor placed close behind the ear and it is then passed through the microphone. The advance version of NAM microphone as shown in Figure 5.2, namely, Open Condenser Wrapped with Soft Silicone (OCWSS) type can record frequency up to 4 kHz [20].

Figure 5.3 shows an example of a NAM signal and its corresponding spectrogram recorded using the NAM microphone. Since the body tissue acts as a lowpass filter, high-frequency information is attenuated, which is clearly seen in a spectrogram of Figure 5.3. In addition, the power of the NAM signal is very small and hence, it cannot be heard by anyone around the speaker. Furthermore, due to recording at the soft body tissue, it is robust to the external noise compared to the normal speech or whispered speech. Hence, the use of this alternate type of communication will indeed help the patient suffering from vocal fold-related diseases.

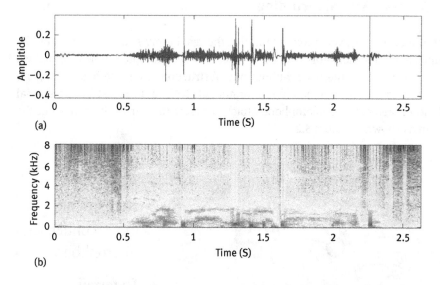

Figure 5.3: (a) Time-domain NAM signal and (b) its spectrogram for an utterance, /"we must learn from the past"/.

5.3.2 Whispered speech signal

During normal speech production, air comes from the lungs and it will cause vocal folds to vibrate; as a result, this excitation will resonate the vocal tract and normal phonated speech will be produced [41]. On the other hand, during whispered speech production, the glottis is opened (i.e., no vocal fold vibrations), which causes exhaled air to pass through the glottal constrictions, which results in turbulent noisy excitation source for vocal tract system and it produces the whispered speech [42].

The whispered speech signal's time-domain waveform and its corresponding spectrogram are shown in Figure 5.4. The whispered speech is more sensitive to the external noise due to its recording at the lips compared to the NAM signal.

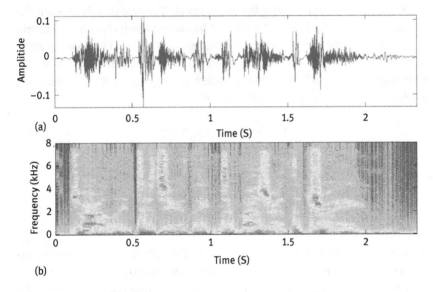

Figure 5.4: (a) Time-domain whispered signal and (b) its spectrogram for an utterance /Charlie did you think to measure the tree?/.

5.4 State-of-the-art conversion techniques

There are very less conversion techniques explored in the area such as NAM-to-normal speech conversion and whisper-to-normal speech conversion [19]. In particular, we have presented here GMM and neural network (NN)-based conversion techniques.

5.4.1 GMM-based conversion technique

Joint density (JD) GMM-based conversion technique finds the mapping function between the source and the target feature vectors. In the case of NAM2WHSP conversion, source and target are the features (spectral or excitation) corresponding to the NAM and the whispered signals, respectively. Similarly, in the case of WHSP2SPCH conversion, source and target are the features corresponding to the whispered and the normal speech signals, respectively. Let the $\mathbf{X} = [\mathbf{x}_1, \mathbf{x}_2, ..., \mathbf{x}_N]$ and $\mathbf{Y} = [\mathbf{y}_1, \mathbf{y}_2, ..., \mathbf{y}_K]$ be the spectral feature vectors corresponding to the source and target, respectively, where $\mathbf{x}_n \in R^d$ and $\mathbf{y}_n \in R^d$ (i.e., $\mathbf{x}$ and $\mathbf{y}$ are d-dimensional feature vectors). The first task is to align the spectral features. Since in the literature, it has been shown that alignment will affect the quality of converted features [43–45]. The joint vector, $\mathbf{Z} = [\mathbf{z}_1, \mathbf{z}_2, ..., \mathbf{z}_r, ..., \mathbf{z}_T]$, is formed after aligning the training database using dynamic time warping (DTW) algorithm, where $\mathbf{z}_r = [\mathbf{x}_n^T, \mathbf{y}_m^T]^T \in R^{2d}$. Furthermore, the joint vector is modeled by a GMM, that is,

$$P(\mathbf{Z}) = \sum_{m=1}^{M} \omega_m^{(\mathbf{z})} N(\mathbf{Z}|\mu_m^{(\mathbf{z})}, \Sigma_m^{(\mathbf{z})}), \tag{5.1}$$

where

$$\mu_m^{(\mathbf{z})} = \begin{bmatrix} \mu_m^{(\mathbf{x})} \\ \mu_m^{(\mathbf{y})} \end{bmatrix} \text{ and } \Sigma_m^{(\mathbf{z})} = \begin{bmatrix} \Sigma_m^{(\mathbf{xx})} & \Sigma_m^{(\mathbf{xy})} \\ \Sigma_m^{(\mathbf{yx})} & \Sigma_m^{(\mathbf{yy})} \end{bmatrix}$$

are the mean vector and covariance matrix of the mth mixture component, respectively, and ω_m is the weight associated with the mth mixture component with $\Sigma_{m=1}^{M} \omega_m^{(\mathbf{z})} = 1$ (constraint for total probability). During training of GMM, the model parameters $\lambda^{(\mathbf{z})} = \{w_m^{(\mathbf{z})}, \mu_m^{(\mathbf{z})}, \Sigma_m^{(\mathbf{z})}\}$ are estimated using the expectation–maximization algorithm [46] (Figure 5.5).

Once the training is completed, the *mapping* function (i.e., $F(\cdot)$) is learned either using the minimum mean squared error (MMSE) criteria [47] or using maximum likelihood estimation (MLE)-based criteria [48]. Since the conditional expectation is the MMSE estimator [49], the mapping function $F(\cdot)$ is given by

$$\hat{\mathbf{y}}_t = F(\mathbf{x}_t) = E[\mathbf{y}_t|\mathbf{x}_t],$$

$$= \int P(\mathbf{y}_t|\mathbf{x}_t, \lambda)\mathbf{y}_t d\mathbf{y}_t,$$

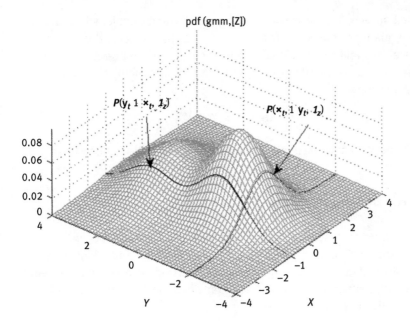

pdf (gmm,[Z])

$P(y_t 1|x_t, 1_z)$

$P(x_t 1|y_t, 1_z)$

Figure 5.5: Representation of conditional pdfs in the JDGMM.

$$= \int \sum_{k=1}^{m} P(k|\mathbf{x}_t, \lambda^z) P(\mathbf{y}_t|\mathbf{x}_t, k, \lambda^z) \mathbf{y}_t d\mathbf{y}_t,$$

$$\hat{\mathbf{y}}_t = \sum_{k=1}^{m} P(k|\mathbf{x}_t, \lambda^z)(E_{mt}^{(\mathbf{y})}).$$

When two variables are jointly Gaussian, their conditional probability distribution function (pdf) (i.e., $P(\mathbf{y}_t|\mathbf{x}_t, \lambda_z)$) will also be Gaussian (as shown in Figure 5.1) with the mean, $E_{mt}^{(\mathbf{y})} = \mu_m^{(\mathbf{y})} + \Sigma_m^{(\mathbf{yx})}(\Sigma_m^{(\mathbf{xx})})^{-1}(\mathbf{x} - \mu_m^{(\mathbf{x})})$, and the covariance matrix, $D_{mt}^{(\mathbf{y})} = \Sigma_m^{(\mathbf{yy})} - \Sigma_m^{(\mathbf{yx})}(\Sigma_m^{(\mathbf{xx})})^{-1}\Sigma_m^{(\mathbf{xy})}$ [49]:

$$\hat{\mathbf{y}} = F(\mathbf{x}) = \sum_{m=1}^{M} p_m(\mathbf{x})(\mu_m^{(\mathbf{y})} + \Sigma_m^{(\mathbf{yx})}(\Sigma_m^{(\mathbf{xx})})^{-1}(\mathbf{x} - \mu_m^{(\mathbf{x})})), \qquad (5.2)$$

where

$$p_m(\mathbf{x}) = P(k|\mathbf{x}_t, \lambda^z) = \frac{\omega_m N(\mathbf{x}|\mu_m^{\mathbf{x}}, \Sigma_m^{\mathbf{xx}})}{\sum_{k=1}^{M} \omega_k N(\mathbf{x}|\mu_k^{\mathbf{x}}, \Sigma_k^{\mathbf{xx}})}$$

is the posterior probability of the source vector $\mathbf{x}$ for the mth mixture component. Statistical averaging in eq. (5.2) leads to the *oversmoothing* of the converted speech. The *oversmoothing* is generally undesirable, which deteriorates the quality of the converted speech signal. To overcome this, use of dynamic features and use of global variance have been proposed to use along with the MLE-based JDGMM method. In the MLE-based method, converted features are given by

$$\hat{\mathbf{y}} = \arg\max_\lambda p(\mathbf{Y}|\mathbf{X}, \lambda^\mathbf{z}), \tag{5.3}$$

where $\mathbf{X}$ and $\mathbf{Y}$ include the dynamic features along with the static features.

5.4.2 NN-based techniques

The relation between spectral feature vectors $\mathbf{X}$ and $\mathbf{Y}$ are obtained using ANN, which consists of K multiple layers [50, 51]. The first, last and middle layers of the ANN are called as input, output and hidden layers, respectively. Here, each layer performs either nonlinear or linear transformation. The transformation at ith layer is given by [52]

$$\mathbf{h}_{i+1} = f(\mathbf{W}_i^T \mathbf{h}_i + \mathbf{b}_i), \tag{5.4}$$

where $\mathbf{h}_i$, $\mathbf{h}_{i+1}$, $\mathbf{W}_i$, $\mathbf{b}_i$ are called as input, output, weights and bias of ith layers, and f is an activation function, which is generally nonlinear (such as tangent hyperbolic, sigmoid and ReLU linear units) or linear. The input and output layers of the ANN are $\mathbf{h}_1 = \mathbf{X}$ and $\mathbf{h}_{K+1} = \mathbf{Y}$. Stochastic gradient descent (SGD) algorithm is used to train the weights and biases of the ANN such that MSE, that is, $E = ||\mathbf{Y} - \hat{\mathbf{Y}}||^2$ is minimized. Figure 5.6 shows the ANN architecture which is used in this chapter.

The ANN with more than two hidden layers (i.e., $K > 2$) is called DNN [53]. The DNN captures the more complex relations between the source and the target spectral feature vectors due to more number of the hidden layers. However, as the number of layers increases, the number of parameters to be estimated also increases. The random initialization of weights and biases of DNNs results in poor convergence, that is, the likelihood will be stuck into local minimum [54]. Hence, we also propose to use nonlinear activation function such as ReLU [55]. Since ReLU converges faster, it is easier to compute. In addition, the effectiveness of ReLU over other nonlinearities has already been discussed in numerous speech applications [25]. Hence, we propose to use ReLU as a nonlinear activation function in our DNN-based conversion technique.

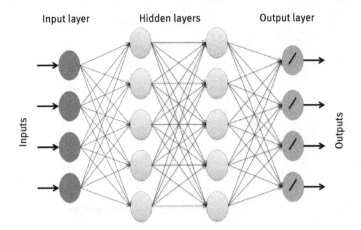

Figure 5.6: The architecture of ANN-based VC framework. Neurons with the slanted line are having linear activation function, and rests are having a nonlinear activation function. After [50].

5.5 Proposed framework

5.5.1 NAM-to-whisper (NAM2WHSP) conversion

The block diagram of the general NAM2WHSP conversion system is shown in Figure 5.7. First, the spectral features are extracted from the NAM signal and whispered speech signal. Since the recording of the NAM and whispered speech has done simultaneously, here, we do not require any alignment before learning the mapping function. In this chapter, we have applied three different methods to learn the mapping function, such as ANN-, GMM- and DNN-based conversion techniques. As both NAM and whispered speech are unvoiced sounds, we did not apply any F_0 conversion techniques. At the time of testing, we extracted the spectral features from the input NAM signal and converted it using learned mapping function. In the end, features are converted into the whispered speech signal using a vocoder.

5.5.2 Whisper-to-speech (WHSP2SPCH) conversion

Due to the unavailability of normal speech from the same speaker from whom the recording of the NAM signal and its corresponding whispered speech is available, we propose to develop a speaker-independent WHSP2SPCH conversion system.

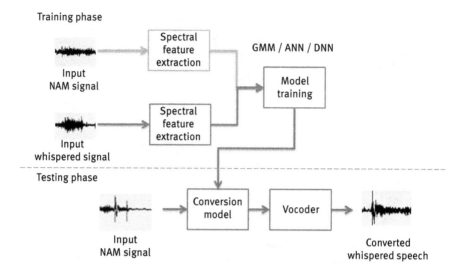

Figure 5.7: Block diagram of the NAM2WHSP conversion system. After [21, 24].

We use multispeakers' whispered speech, and its corresponding normal speech databases to develop our proposed speaker-independent WHSP2SPCH conversion system. Spectral conversion techniques are almost the same as mentioned in the NAM2WHSP conversion. However, one of the main issues before learning the mapping function for the WHSP2SPCH conversion system is that whisper and its corresponding normal speech signals are not aligned. To align whisper and normal speech, researchers have used DTW algorithm and also force-aligned labels. In this chapter, we used the DTW algorithm to align spectral and excitation features of whisper and its corresponding speech signal. We used DNN-based conversion method to learn the mapping function of spectral feature conversion due to the availability of a large amount of data from multispeakers. In the earlier work, researchers have tried GMM- and ANN-based method due to the limited amount of data available for development of the speaker-dependent WHSP2SPCH conversion system. In order to predict F_0 from the whispered speech, we used the following three different approaches:

- Predict F_0 jointly with the spectral features [21]
- Predict F_0 directly from whisper spectral features [22]
- Predict F_0 separately from the converted spectral features after WHSP2SPCH conversion [26]

In order to predict jointly F_0 and the spectral features of the natural speech, we concatenate target (i.e., normal speech) cepstral features with 1D F_0 value for that frame. Hence, our joint vector $Z = [X; Y; F_0]^T$ will be having $(2D + 1)$ dimensions, where $x, y \in \Re^d$ and $F_0 \in \Re^1$. In the second approach, We trained the separate mapping function for predicting F_0 values for directly predicting F_0 from the cepstral features corresponding to the whispered speech. Here, we considered only the voiced frames for learning the mapping function for F_0 prediction. Hence, a voiced–unvoiced detector (VUD) has been used. To detect voiced or unvoiced frames, we only used the ANN-based model. Since the number of voiced frames after the alignment will be very less compared to all the frames, we did not use the DNN-based approach for VUD system. Similarly, in the third approach, instead of predicting F_0 from the whispered speech, we used converted cepstral features for VUD decision as well as for learning mapping function for F_0. During the conversion, the given spectral features from the whispered speech are converted using the learned mapping function. In addition, based on the VUD decision from the spectral features, F_0 values are predicted for voiced frames using learned mapping function for F_0 prediction. The block diagram of the proposed WHSP2SPCH conversion system is shown in Figure 5.8.

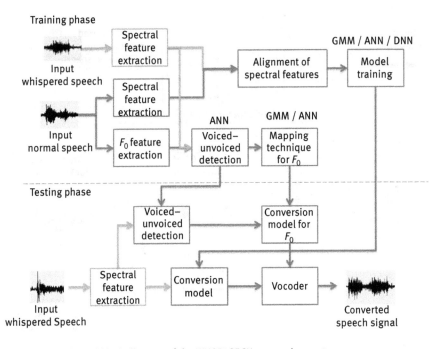

Figure 5.8: Proposed block diagram of the WHSP2SPCH conversion system.

5.5.3 NAM-to-normal speech conversion (NAM2SPCH)

Once both the NAM2WHSP and the WHSP2SPCH systems are trained, we combined both the systems to develop the proposed NAM2SPCH conversion system. The basic block diagram of the proposed NAM2SPCH conversion system is shown in Figure 5.9. Here, from a given NAM signal, first, we extract spectral features corresponding to the NAM signal, and then converted it using the NAM2WHSP system. These converted cepstral features are then passed through VUD and for voiced frames, the F_0 values are predicted. On the other hand, the converted cepstral features obtained after the NAM2WHSP systems are further passed through the speaker-independent WHSP2SPCH conversion system, and then converted corresponding to the normal speech signal's cepstral features. In the end, both these cepstral features and its predicted F_0 values are passed through the VOCODER and these features are converted into the audible normal speech signal. Figure 5.10(a) and (b) shows the time-domain waveform and spectrogram for an original NAM signal and Figure 5.10(c) and (d) shows the time-domain waveform and spectrogram for the converted speech signal. The high-frequency information absent in the NAM signal is extracted and presents

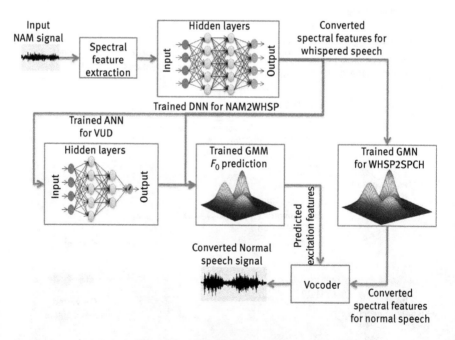

Figure 5.9: Proposed block diagram for NAM2SPCH conversion system.

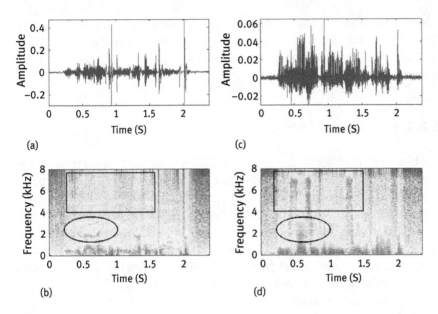

Figure 5.10: (a) Original NAM signal, (b) its corresponding spectrogram, (c) converted speech signal from NAM and (d) its spectrogram for an utterance /Noone has yet been charged./.

in the converted speech signal, which is shown by a rectangle box in Figure 5.10. In addition, harmonics that are absent in the NAM signal due to the absence of vocal fold vibration is recovered in the converted speech signal as shown via ellipse.

5.6 Experimental setup and results

5.6.1 Databases

In this chapter, the CSTR NAM TIMIT Plus corpus has been used for the development of NAM2WSHP conversion systems [56]. The corpus contains 420 Herald Glasgow newspaper texts, 460 TIMIT texts and 18 isolated command words. We have used total 420 utterances of whispered speech and their corresponding NAM signals from Herald text. Out of these 420 utterances, we have selected 400 utterances for the training and the remaining 20 utterances for the evaluation.

In order to develop speaker-independent WHSP2SPCH conversion systems, we have used total 40 speakers' whisper and their corresponding normal speech data from the two databases, namely, the CHAracterizing INdividual Speakers (CHAINS) Speech Corpus [57] and the EMG-UKA Trail corpus [58]. We have taken in total 1,302 utterances for the training and 126 utterances for the evaluation.

5.6.2 Experimental setup

For both the NAM2WHSP and the WHSP2SPCH conversion systems, 25-D mel cepstral coefficients (MCCs) (including the 0th coefficient) and 1-D F_0 per frame (with 25 ms frame duration and 5 ms frameshift) have been extracted using AHOCODER analysis–synthesis method [59]. In particular, NAM2WHSP conversion systems have been developed using JDGMM-based, ANN-based and DNN-based methods. For JDGMM-based NAM2WHSP systems, the number of mixture components has been varied from $m = 8, 16, 32, 64$ and 128. For ANN-based systems, we used two hidden layers NN with ReLU and sigmoid nonlinear activation functions. The number of neurons is 25, 50, 50 and 25 in the input layer, two hidden layers and output layer, respectively. Similarly, proposed DNN-based NN consists of three hidden layers. The number of neurons in the DNN-based networks is 25, 512, 512, 512 and 25 in the input layer, hidden layers and output layer, respectively. We also explored sigmoid and ReLU nonlinear activation functions in DNN. For training of both ANN and DNN, we used SGD algorithm. About 200 number of iterations were performed during training with mini-batch size of 150. The learning rate, momentum and decay rate were set to 0.01, 0.3 and 5×10^{-5}, respectively. Furthermore, we also explored the similar architectures mentioned earlier with the same details for the WHSP2SPCH conversion systems.

5.6.3 Objective analysis

We have applied appropriate objective measure based on the objective of the system. In particular, we have applied mel cepstral distortion (MCD) and perceptual evaluation of speech quality (PESQ) to measure the effectiveness of the NAM2WHSP conversion systems [60]. On the other hand, for WHSP2SPCH conversion systems, we have considered the MCD as well as F_0-based objective measures [61], such as log root mean square error (RMSE) and voice–unvoiced detection (VUD) accuracy, since the main goal is to accurately predict F_0 from unvoiced whispered speech in the WHSP2SPCH system. The traditional MCD measure is used here which is given by [48]

$$\text{MCD [dB]} = \frac{10}{\ln 10} \sqrt{2 \sum_{i=1}^{25} (m_i^t - m_i^c)^2}, \tag{5.5}$$

where m_i^t and m_i^c are the ith MCCs of the reference, and converted signal's cepstral features. Here, target signal's cepstral features will be corresponding to the whispered speech in the NAM2WHSP. Similarly, target signal's cepstral features will be corresponding to the normal speech in the WHSP2SPCH system. MCD is the distance between converted and reference cepstral features. Hence, the systems that are having lesser MCD values are considered as good systems. The PESQ is a very well-known objective measure to evaluate the quality of speech. It is calculated by taking the linear combination of d_{sym} (average normal disturbance value) and d_{asym} (average assymetrical disturbance value) between reference and converted signals, which is given by [60]

$$\text{PESQ} = 4.5 - 0.1 \times d_{\text{sym}} - 0.0309 \times d_{\text{asym}}. \tag{5.6}$$

The system that is having high PESQ score is considered as the good system. Table 5.1 shows the results corresponding to the objective measure for the developed various NAM2WHSP conversion systems. It is clearly visible that our proposed DNN-based systems working better compared to the ANN, and the GMM-based systems in the context of both the MCD and the PESQ.

Table 5.1: Objective analysis of the NAM2WHSP systems.

	ANN		DNN		GMM				
	ReLU	Sigmoid	ReLU	Sigmoid	$m = 8$	$m = 16$	$m = 32$	$m = 64$	$m = 128$
MCD (dB)	3.30	3.32	**3.26**	3.42	3.28	3.27	3.26	3.26	3.26
PESQ	2.30	2.32	**2.41**	1.97	2.32	2.37	2.38	2.40	2.40

m represents the number of mixtures.

Table 5.2 shows the MCD for various WHSP2SPCH-based conversion systems. From our analysis, we have found that GMM with $m = 128$ mixture components works better compared to all other systems. To measure the log(RMSE) of F_0, first the converted speech signal and then the actual reference signal are time-aligned using DTW algorithm. The DTW-aligned pairs will generate voiced–voiced, voiced–unvoiced, unvoiced–voiced and unvoiced–unvoiced pairs. Here, we consider voiced–voiced pairs for computing the log(RMSE) of the F_0 value, which is given by

Table 5.2: MCD analysis of the proposed speaker-independent WHSP2SPCH systems.

	ANN		DNN		GMM				
	ReLU	Sigmoid	ReLU	Sigmoid	$m = 16$	$m = 32$	$m = 64$	$m = 128$	$m = 256$
MCD (dB)	8.06	8.26	7.86	8.34	7.83	7.78	7.77	7.74	7.77

m represents the number of mixtures

$$\log(\text{RMSE}) = \log\left(\sqrt{\sum_{i=1}^{k} (F_{0i}^{t} - F_{0i}^{c})^2}\right), \tag{5.7}$$

where k is the total number of voiced–voiced pairs after alignment, and F_0^t and F_0^c are the fundamental frequencies (F_0) corresponding to the reference and the converted speech signals, respectively. The lesser the value of the log (RMSE), better the system is. VUD accuracy is defined as

$$\text{VUD} = \frac{\text{VV} + \text{UU}}{N}, \tag{5.8}$$

where VV is the total number of voiced–voiced pairs, UU is the total number of unvoiced–unvoiced pairs and N is the total number of pairs after DTW alignment. The system that is having the more VUD accuracy is considered as the best.

Table 5.3 shows the objective evaluation for F_0 estimated jointly with cepstral features. Table 5.4 shows the objective evaluations for F_0 estimated from the converted cepstral coefficients. Table 5.5 shows the objective evaluation for F_0 estimated directly from the cepstral coefficients of the whisper. It can be observed that log RMSE is least when F_0 is estimated using GMM-based conversion technique with 32 number of mixtures. However, accuracy of VUD is more for the ANN-based system when it is estimated directly from the cepstral coefficients of whisper. From objective evaluation of the NAM2WHSP and the WHSP2SPCH conversion systems, we have selected various components of the NAM2SPCH conversion systems based on their performance.

Table 5.3: Objective analysis of F_0 estimated jointly in WHSP2SPCH systems.

	ANN		DNN		GMM			
	ReLU	Sigmoid	ReLU	Sigmoid	$m = 16$	$m = 32$	$m = 64$	$m = 128$
log(RMSE)	10.39	9.88	10.62	8.69	11.60	10.96	10.84	10.30
VUD accuracy (%)	57.28	54.24	60.97	52.47	63.65	61.51	61.31	60.32

m represents the number of mixtures.

Table 5.4: Objective analysis of F_0 estimated from WHSP2SPCH converted cepstrum.

	ANN		DNN		GMM			
	ReLU	Sigmoid	ReLU	Sigmoid	$m = 16$	$m = 32$	$m = 64$	$m = 128$
log(RMSE)	5.86	6.96	6.27	6.53	4.43	**4.37**	4.43	4.40
VUD Accuracy (%)	62.64	68.09	65.62	65.70	58.55	59.65	59.81	59.13

m represents the number of mixtures.

Table 5.5: Objective analysis of F_0 estimated from cepstral coefficients of the whispered speech.

	ANN	GMM			
	ReLU	$m = 16$	$m = 32$	$m = 64$	$m = 128$
log RMSE	6.51	5.97	5.91	5.90	5.89
VUD accuracy (%)	**74.83**	74.59	74.50	74.41	73.856

m represents the number of mixtures.

Figure 5.11 presents the $\log(F_0)$ predicted using GMM-based WHSP2SPCH system with $m = 32$. In addition, for VUD, ANN-based architecture was selected. We can observe that the length of the F_0 contour for the converted signal and its corresponding natural signal is different.

5.6.4 Subjective analysis

The main aim of this chapter is to extract the message recorded via NAM microphone. Hence, we have focused on various intelligibility tests for subjective evaluations of various proposed systems. For subjective evaluations, we have considered three tests, namely, ABX test, word error rate (WER) and mean opinion scores (MOS) test for the intelligibility [62]. Total 15 listeners (12 males and 3 females with age between 21 to 29 years) took part in all the subjective tests. We used high-quality BOSE headphones for subjective evaluations. In the WER test, we asked subjects to transcribe six randomly played utterances from the three different systems. We told not to replay utterance more than 4 times (in order to avoid cognitive-related bias in hearing and subjective judgment) during their transcription. Based on their submission, % WER is calculated as [63]

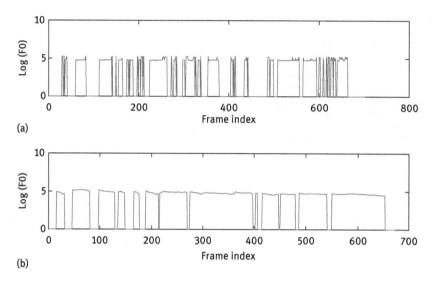

(a)

(b)

Figure 5.11: For an utterance, / " The same shelter could be built into an embankment or below ground level "/ (a) log (F_0) predicted using GMM-based WHSP2SPCH system and (b) corresponding log (F_0) of natural speech signal.

$$\mathrm{WER}(\%) = \frac{I + D + S}{T} \times 100, \tag{5.9}$$

where I, D and S represent the number of insertions, deletions and substitutions, respectively, and T is the total number of words in a given utterance.

From Table 5.6, it is shown that % WER of NAM is very high, that is, hardly one word is intelligible. However, NAM2WHSP and NAM2SPCH systems have very less % WER compared to the NAM. Hence, our goal of extracting the linguistic message from the NAM signal is indeed achieved. In particular, the NAM2SPCH system is having less % WER compared to the NAM2WHSP conversion system. However, the average number of times the subjects replayed the utterances is more compared to the NAM2WHSP system.

We have taken four utterances from each system to evaluate the MOS test for intelligibility. Subjects were asked to rate each randomly played utterances based on the intelligibility from the different systems on a five-point scale (1 = not at all intelligible; 2 = hardly one or two words are intelligible; 3 = half of the message is intelligible; 4 = mostly intelligible and 5 = completely intelligible). The analysis of MOS score along with 95% confidence interval is shown in Figure 5.12, which illustrates that NAM2WHSP is more intelligible compared to the NAM2SPCH systems. Furthermore, it can be seen that DNN-based

Table 5.6: % WER for the developed systems.

	NAM	NAM2WHSP	NAM2SPCH
WER (%)	81	31.5	27.35
Number of replays	2.89	2.52	2.57

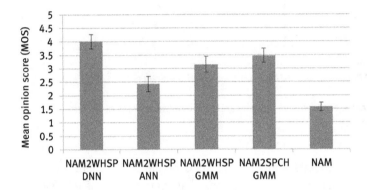

Figure 5.12: MOS analysis for intelligibility of various systems along with 95% confidence interval.

NAM2WHSP system performs better than the ANN- or GMM-based NAM2SPCH systems. Moreover, poor MOS score of NAM signal also indicates that the NAM signal is not at all intelligible. We have also taken ABX test for intelligibility between NAM2WHSP and NAM2SPCH systems. Subjects were asked to prefer the utterance that is more intelligible. The result of the ABX test is shown in Figure 5.13. We can see that the subjects have preferred NAM2SPCH system compared to the NAM2WHSP system.

5.7 Summary and conclusions

In this chapter, we presented an overview of NAM2WHSP or NAM2SPCH conversion system. We also discussed the medical relevance of the proposed research. In particular, we propose DNN-based NAM2WHSP and NAM2SPCH conversion systems with ReLU as the nonlinear activation function. We compared our results with the state-of-the-art GMM-based conversion system. Furthermore, we also compared the performance of the DNN-based system with various

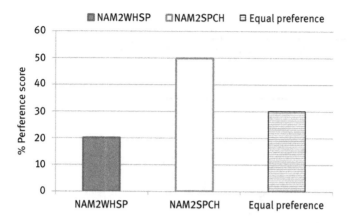

Figure 5.13: ABX test analysis for intelligibility preference between NAM2WHSP versus NAM2SPCH system.

nonlinear activation functions against the ReLU nonlinear activation function. We observed from the objective evaluations that there is 25% of relative improvement of DNN-based NAM2WHSP system than the baseline GMM-based NAM2WHSP system. In addition, due to unavailability of the normal speech utterances from the same speaker whose NAM and whispered speech is available we proposed to develop a speaker-independent WHSP2SPCH conversion system. The proposed WHSP2SPCH leads us to 4.15% of the absolute decrease in WER. Moreover, from the ABX test, we have found that our proposed NAM2SPCH system is 30% of times more preferred than the NAM2WHSP system in terms of intelligibility. We found that both the NAM2WHSP and NAM2SPCH systems are able to extract linguistic message very well from the NAM signal. However, we observed the poor quality of F_0 estimation, which deteriorates the naturalness of the converted speech signal in the NAM2SPCH system. Hence, our future research work will be directed toward improving the naturalness of the NAM2SPCH converted voice.

Acknowledgments: The authors would like to thank the Ministry of Electronics and Information Technology (MeitY), Govt. of India, for consortium-sponsored project development of Text-To-Speech (TTS) Synthesis System in Indian Languages (Phase-II) and authorities of DA-IICT, Gandhinagar, for their kind support and cooperation to carry out this research. We thank all the participants who took part in the subjective evaluations during this research work.

References

[1] Rubin, A. D., and Sataloff, R. T. "Vocal fold paresis and paralysis,". Otolaryngologic Clinics of North America, 40(5), 1109–1131, 2007.

[2] Sulica, L. Vocal fold paresis: an evolving clinical concept. Current Otorhinolaryngology Reports, 1(3), 158–162, 2013.

[3] Wallis, L., Jackson-Menaldi, C., Holland, W., and Giraldo, A. Vocal fold nodule vocal fold polyp: answer from surgical pathologist and voice pathologist point of view. Journal of Voice, 18(1), 125–129, 2004.

[4] Mattiske, J. A., Oates, J. M., and Greenwood, K. M. Vocal problems among teachers: a review of prevalence, causes, prevention, and treatment. Journal of Voice, 12(4), 489–499, 1998.

[5] Kresten, P. V. Silent Speech Interface, 1em plus 0.5em minus 0.4em. Volut Press, 2012.

[6] Nakajima, Y., Kashioka, H., Shikano, K., and Campbell, N. "Non-audible murmur recognition input interface using stethoscopic microphone attached to the skin," in IEEE International Conference on Acoustics, Speech, and Signal Processing (ICASSP), Hong Kong, 2003, pp. 704–708.

[7] Denby, B., Schultz, T., Honda, K., Hueber, T., Gilbert, J. M., and Brumberg, J. S. Silent speech interfaces. Speech Communication, 52(4), 270–287, 2010.

[8] Kaburagi, T., Wakamiya, K., and Honda, M. Three-dimensional electromagnetic articulography: A measurement principle. The J. of the Acoust. Soc. of Amer. (JASA), 118 (1), 428–443, 2005.

[9] Jou, S.-C. S., Schultz, T., and Waibel, A. Adaptation for soft whisper recognition using a throat microphone, in INTERSPEECH, Jeju Island, Korea, 2004, 1493–1496.

[10] Schultz, T., and Wand, M. Modeling coarticulation in EMG-based continuous speech recognition. Speech Communication, 52(4), 341–353, 2010.

[11] Wand, M., Schulte, C., Janke, M., and Schultz, T. Array-based electromyographic silent speech interface, in BIOSIGNALS, Barcelona, Spain, 2013, 89–96.

[12] Jorgensen, C., and Binsted, K. "Web browser control using EMG based sub vocal speech recognition," in Proceedings of the Annual Hawaii International Conference on System Sciences (HICSS), Big Island, HI, USA, 2005, pp. 1–4.

[13] Jou, S.-C., Schultz, T., Walliczek, M., Kraft, F., and Waibel, A. Towards continuous speech recognition using surface electromyography, in INTERSPEECH, Pittsburgh, USA, 2006, 573–576.

[14] Hueber, T., Aversano, G., Cholle, G., Denby, B., Dreyfus, G., Oussar, Y., Roussel, P., and Stone, M. "Eigentongue feature extraction for an ultrasound-based silent speech interface," in IEEE International Conference on Acoustics, Speech and Signal Processing (ICASSP), vol. 1, Honolulu, Hawaii, USA, 2007, pp. 1240–1245.

[15] Hueber, T., Chollet, G., Denby, B., Dreyfus, G., and Stone, M. Phone recognition from ultrasound and optical video sequences for a silent speech interface, in INTERSPEECH, Brisbane, Australia, 2008, 1–4.

[16] Hueber, T., Benaroya, E.-L., Chollet, G., Denby, B., Dreyfus, G., and Stone, M. Development of a silent speech interface driven by ultrasound and optical images of the tongue and lips. Speech Communication, 52(4), 288–300, 2010.

[17] Toda, T., Nakamura, K., Sekimoto, H., and Shikano, K. "Voice conversion for various types of body transmitted speech," in IEEE International Conference on Acoustics, Speech and Signal Processing (ICASSP), Taipei, Taiwan, 2009, pp. 3601–3604.

[18] Tajiri, Y., Kameoka, H., and Toda, T. "A noise suppression method for body-conducted soft speech based on non-negative tensor factorization of air-and body-conducted signals," in IEEE International Conference on Acoustics, Speech and Signal Processing (ICASSP), New Orleans, USA, 2017, pp. 4960–4964.

[19] Toda, T., Nakagiri, M., and Shikano, K. Statistical voice conversion techniques for body-conducted unvoiced speech enhancement. IEEE Transactions on Audio, Speech, and Language Processing, 20(9), 2505–2517, 2012.

[20] Tajiri, Y., Tanaka, K., Toda, T., Neubig, G., Sakti, S., and Nakamura, S. Non-audible murmur enhancement based on statistical conversion using air-and body-conductive microphones in noisy environments, in INTERSPEECH, Dresden, Germany, 2015, 2769–2773.

[21] Toda, T., and Shikano, K. NAM-to-speech conversion with Gaussian mixture models, in INTERSPEECH, Lisbon, Portugal, 2005, 1957–1960.

[22] Tran, V.-A., Bailly, G., Lœvenbruck, H., and Toda, T. Multimodal HMM-based NAM-to-speech conversion, in INTERSPEECH, Brighton, United Kingdom (UK), 2009, 656–659.

[23] Tran, V.-A., Bailly, G., Lœvenbruck, H., and Toda, T. Improvement to a NAM-captured whisper-to-speech system. Speech Communication, 52(4), 314–326, 2010.

[24] Shah, N., Shah, N. J., and Patil, H. A. Effectiveness of generative adversarial network for non-audible murmur-to-whisper speech conversion, in INTERSPEECH, Hyderabad, India, 2018, 3157–3161.

[25] Zeiler, M. D., Ranzato, M., Monga, R., Mao, M., Yang, K., Le, Q. V., Nguyen, P. Senior, A., Vanhoucke, V., Dean, J. et al. "On rectified linear units for speech processing," in IEEE International Conference on Acoustics, Speech and Signal Processing (ICASSP), Vancouver, British Columbia, Canada, 2013, pp. 3517–3521.

[26] Janke, M., Wand, M., Heistermann, T., Schultz, T., and Prahallad, K. "Fundamental frequency generation for whisper-to-audible speech conversion," in IEEE International Conference on Acoustics, Speech and Signal Processing (ICASSP), Florence, Italy, 2014, pp. 2579–2583.

[27] Sharifzadeh, H. R., McLoughlin, I. V., and Ahmadi, F. Reconstruction of normal sounding speech for laryngectomy patients through a modified celp codec. IEEE Transactions on Biomedical Engineering, 57(10), 2448–2458, 2010.

[28] Morris, R. W., and Clements, M. A. Reconstruction of speech from whispers. Medical Engineering & Physics, 24(7), 515–520, 2002.

[29] McLoughlin, I. V., Li, J., and Song, Y. Reconstruction of continuous voiced speech from whispers, in INTERSPEECH, Lyon, France, 2013, 1022–1026.

[30] Mcloughlin, I. V., Sharifzadeh, H. R., Tan, S. L., Li, J., and Song, Y. Reconstruction of phonated speech from whispers using formant-derived plausible pitch modulation. ACM Transactions on Accessible Computing (TACCESS), 6(4), 12, 2015.

[31] Shah, N. J., Parmar, M., Shah, N., and Patil, H. A. "Novel MMSE DiscoGAN for cross-domain whisper-to-speech conversion," in Machine Learning in Speech and Language Processing (MLSLP) Workshop, Google Office, Hyderabad, India, 2018.

[32] Aronson, A. E., and Bless, D. Clinical Voice Disorders, Thieme, 2011.

[33] Wu, A. P., and Sulica, L. Diagnosis of vocal fold paresis: current opinion and practice. The Laryngoscope, 125(4), 904–908, 2015.

[34] Simpson, B., and Rosen, C. Glottic insufficiency: Vocal fold paralysis, paresis, and atrophy. Operative Techniques in Laryngology, 29–35, 2008.

[35] "Vocal cord paralysis," URL: https://www.asha.org/public/speech/disorders/vfparaly sis/, {Last Accessed: October 15, 2017}.

[36] Jacob, P., Kahrilas, P., and Herzon, G. Proximal esophageal ph-metry in patients with reflux laryngitis. Gastroenterology, 100(2), 305–310, 1991.

[37] "Vocal cord nodule and polyp," URL: https://www.asha.org/public/speech/disorders/ NodulesPolyps/, {Last Accessed: October 15, 2017}.

[38] "Vocal abuse and misuse," URL: https://www.timetohear.com/vocal-abuse/, {Last Accessed: October 15, 2017}.

[39] "Vocal fold paralysis and vocal fold lsion," URL: https://www.advancedreconstruction. com/stroke-treatment/vocal-cord-paralysis/, {Last Accessed: October 15, 2017}.

[40] Toda, T., Nakamura, K., Nagai, T., Kaino, T., Nakajima, Y., and Shikano, K. Technologies for processing body-conducted speech detected with non-audible murmur microphone, in INTERSPEECH, Brighton, United Kingdom, 2009, 632–635.

[41] Quatieri, T. F. Discrete Time Speech Signal Processing: Principles and Practice, Pearson Education India, 2006.

[42] Tartter, V. C. What is in a whisper. The J. of the Acoust. Soc. of Amer. (JASA), 86(5), 1678–1683, 1989.

[43] Helander, E., Schwarz, J., Nurminen, J., Silen, H., and Gabbouj, M. On the impact of alignment on voice conversion performance, in INTERSPEECH, Brisbane, Australia, 2008, 1–5.

[44] Rao, S. V., Shah, N. J., and Patil, H. A. Novel pre-processing using outlier removal in voice conversion, in 9th ISCA Speech Synthesis Workshop, Sunnyvale, CA, USA, 2016, 147–152.

[45] Shah, N. J., and Patil, H. A. Analysis of features and metrics for alignment in text-dependent voice conversion, in International Conference on Pattern Recognition and Machine Intelligence (PReMI), ISI, Kolkata: B. Uma Shankar et. al., Lecture Notes in Computer Science (LNCS), Springer, Vol. 10597, 2017, 299–307.

[46] Dempster, A. P., Laird, N. M., and Rubin, D. B. Maximum likelihood from incomplete data via the EM algorithm. Journal of the Royal Statistical Society, 39(1), 1–38, 1977.

[47] Kain, A., and Macon, M. W. "Spectral voice conversion for text-to-speech synthesis," in IEEE International Conference on Acoustics, Speech and Signal Processing (ICASSP), Seattle, WA, 1998, pp. 285–288.

[48] Toda, T., Black, A. W., and Tokuda, K. Voice conversion based on maximum-likelihood estimation of spectral parameter trajectory. IEEE Trans. on Audio, Speech and Lang. Process., 15(8), 2222–2235, 2007.

[49] Kay, S. M. Fundamentals of Statistical Signal Processing, Volume I: Estimation Theory, Upper Saddle River, New Jersey: Prentice Hall, 1998.

[50] Desai, S., Raghavendra, E. V., Yegnanarayana, B., Black, A. W., and Prahallad, K. "Voice conversion using artificial neural networks," in IEEE International Conference on Acoustics, Speech and Signal Processing (ICASSP), Taipei, Taiwan, 2009, pp. 3893–3896.

[51] Desai, S., Black, A. W., Yegnanarayana, B., and Prahallad, K. Spectral mapping using artificial neural networks for voice conversion. IEEE Transactions on Audio, Speech, and Language Processing, 18(5), 954–964, 2010.

[52] Yegnanarayana, B. Artificial Neural Network, PHI Learning Pvt. Ltd.,2009.

[53] Goodfellow, I., Bengio, Y., and Courville, A. Deep Learning, The MIT Press,2016.

[54] Glorot, X., and Bengio, Y. "Understanding the difficulty of training deep feedforward neural networks," in International Conference on Artificial Intelligence and Statistics, Sardinia, Italy, 2010, pp. 249–256.

[55] Maas, A. L., Hannun, A. Y., and Ng, A. Y. "Rectifier nonlinearities improve neural network acoustic models," in International Conference on Machine Learning (ICML), vol. 30, no. 1, Atlanta, USA, 2013.

[56] "Publicly available: The CSTR NAM TIMIT Plus Corpus," URL: homepages.inf.ed.ac.uk/jya magis/release/CSTR-NAM-TIMIT-Plus-ver0.81.tar.gz, {Last Accessed: October 15, 2017}.

[57] Grimaldi, M., and Cummins, F. Speaker identification using instantaneous frequencies. IEEE Transactions on Audio, Speech, and Language Processing, 16(6), 1097–1111, 2008.

[58] Wand, M., Janke, M., and Schultz, T. The EMG-UKA corpus for electromyographic speech processing, in INTERSPEECH, Singapore, 2014, 1–5.

[59] Erro, D., Sainz, I., Navas, E., and Hernáez, I. Improved HNM-based vocoder for statistical synthesizers, in INTERSPEECH, Florence, Italy, 2011, 1809–1812.

[60] Hu, Y., and Loizou, P. C. Evaluation of objective quality measures for speech enhancement. IEEE Transactions on Audio, Speech, and Language Processing, 16(1), 229–238, 2008.

[61] Konno, H., Kudo, M., Imai, H., and Sugimoto, M. Whisper-to-normal speech conversion using pitch estimated from spectrum. Speech Communication, 83, 10–20, 2016.

[62] Rec, I. P. 85. a method for subjective performance assessment of the quality of speech voice output devices, International Telecommunication Union (ITU), Geneva., Available Online: https://www.itu.int/rec/T-REC-P.85-199406-I/en. {Last Accessed: September 26, 2017}.

[63] Rabiner, L. R., and Juang, B.-H. Fundamentals of Speech Recognition, PTR Prentice Hall, 1993.

Part III: **Use of novel speech diagnostic and therapeutic intervention software for speech enhancement and rehabilitation**

T.V. Ananthapadmanabha

6 Application of speech signal processing for assessment and treatment of voice and speech disorders

Abstract: A person's holistic health involves not only their physiological and emotional well-being but also one's ability to communicate. The issues addressed in this chapter cover general awareness about communication disabilities and possible technological solutions for an early identification of the communication disorder/disability as well as software-based therapy solutions for improving communication skills. The following topics are covered: (a) an overview of communication disabilities, (b) important parameters of a speech production model, (c) a brief introduction to speech analysis and synthesis as relevant to communication disabilities and (d) an overview of Vagmi and other software solutions developed by the author. The style of presentation is intentionally nontechnical so as to appeal to a more general readership.

Keywords: voice disorders, speech disorders, voice analysis, voice therapy, speech therapy, speech signal processing, Vagmi

6.1 Introduction

This chapter is written in an informal style with a primary focus on the author's own research and product development in order to present a general awareness of the work in the field. It is not an exhaustive review chapter. The author's work has been inspired by and is build on previous work of many researchers in this field.

Speech consists of acoustic waves, a stream of audible sounds, which convey information and emotions. Speech production is a learned skilled activity that involves mind and the brain at the highest level and at the peripheral level, the physiological instruments of speech production, namely, the respiratory, the laryngeal and the articulatory systems. However, given the ease with which most persons abundantly use the speaking modality in everyday life the underlying mechanism of production is often taken for granted. Problems related to speech production may arise at any age, right from birth (congenital) to the old age and

T. V. Ananthapadmanabha, Voice and Speech Systems, Bangalore, India

https://doi.org/10.1515/9781501501265-007

due to various reasons. Changes in speech quality or difficulties in speech production may be either a mere passing event or more ominously a precursor for an impending long-term problem. Speech production problems are broadly categorized as (a) organic disorders such as cleft palate and damage to vocal folds; (b) neurological disorders such as muscular atrophy, amyotrophic lateral sclerosis (ALS), Parkinson's, Alzheimer's and aphasia due to stroke; (c) functional disorders that are deficiencies due to a lack of awareness or learnt bad habits in production. For example, abuse or improper use of voice, mispronunciation and disfluency (mannerism of using fillers like "mmm" or "ok"). These three broad categories (organic, neurological and functional) of problems may affect one or more of the three physiological subsystems (respiratory, laryngeal, and articulatory) forming a matrix of disorders.

Improper use of grammar (tense, number, etc.) is considered to be a language disorder. It may arise due to cognitive impairment. A child is prompted to speak ex-tempore or narrate a story after showing a picture. The narration is recorded and later analyzed for the number and type of grammatical mistakes. There are definite milestones to be achieved in language usage with development in children. Failure to achieve these milestones is considered to be a language disorder. Signal-processing techniques are not used in the assessment or treatment of a language disorder and hence is not addressed in this chapter.

Even a slightest deviation from normal production is perceivable in the auditory quality of speech. Hence, the speech signals carry information about any underlying disorder. Nevertheless, the challenge is to analyze a deviant speech signal using signal-processing techniques to identify the type and degree (extent) of an underlying disorder. Such an effort is termed as the assessment or diagnosis of voice and speech disorders. There are many successful models of speech production at the peripheral level. These models are characterized by measurable acoustic parameters such as the pitch, the level of speaking (intensity), the formant frequencies and the articulatory positions. These parameters can be extracted from a speech signal spoken by a subject, using signal-processing techniques; they can be objectively compared against the normative (the range of values of the parameters for a normal population of the same gender and same age group) in order to ascertain if the voice quality and/or speech production is normal or deviant. Using such objective measurements, a final assessment or interpretation of the underlying deficiency and its degree (moderate to severe) is made by an experienced speech language pathologist. The corrective measures are then planned.

In the case of organic disorders such as cleft palate, surgery is performed to repair the deviant anatomical structure. Subsequently, speech therapy has to be prescribed for the child to correct improper pronunciation habits learnt prior to surgery. For the case of a hard of hearing child, who is fitted either with a hearing aid or with a cochlear implant, speech therapy sessions are prescribed. Such therapy sessions are primarily available in major cities, since children residing in villages cannot commute long distances to undertake therapy sessions. As a result, remote access to computerized therapy via Internet constitutes a possible solution. Otherwise, professionals have to periodically visit the remote villages and conduct camps.

In a remedial computerized voice and speech therapy session, a child with misarticulation or a hard of hearing child using an aid is given a continuous visual feedback of the speech production process currently being used by the child as against a model or a reference to assist the child to learn proper pronunciation and/or intonation. Functional voice disorders arising due to lack of awareness or bad habits may also be corrected by means of computer-assisted therapy sessions. Certain tested techniques are known to be effective in correcting disfluency (stammering). In such cases where computerized speech therapy is inapplicable (e.g., neuromuscular disorder), technology provides augmentative communication devices and assistive pointers. For example, a speech synthesizer or an artificial larynx restores partially the communication ability in order to meet basic survival needs and requirements.

Even persons with the so-called normal voice may have to be concerned about voice production because they may be inadvertently abusing their voice causing damage to the vocal folds. Their voice may become weak or hoarse after a prolonged use if not corrected. At an early stage, such an abuse may be detected and corrected by means of voice therapy. If ignored, surgery of vocal folds may be required at a later date. Professional voice users such as teachers, singers, lawyers and actors may not be fully aware of the potentiality of their voice range. They may be interested in enhancing their voice quality. The tools of voice assessment and enhancement therapy are of relevance to them as well.

6.2 An overview of voice and speech disorders

6.2.1 Importance of language and speech

Language is the code for conveying a message. Some philosophers and neurosurgeons believe that in the evolution of species, a human being's self-consciousness or self-awareness arose because of language and speech. Unlike in animals where there is a fixed repertoire of messages, sounded as an instinctive reaction, in humans speech communication is a voluntary act and the number of messages that can be formed is unlimited. This is made possible by concatenating words in combinatorial ways under a syntactic constraint, specific to a given language. Thus, even with a finite number of words, a large number of sentences may be formed. Similarly, words are formed by a combination of a finite number of (about 50) speech sounds. This is made possible by our perceptual ability to "abstract a speech sound" despite the context (surrounding sounds) in which it appears. Although gestures may be used for conveying messages, speech is the most efficient way of communication since the hands are free for other activities; moreover, line of sight is not a requirement and a faster rate of communication is possible. Speech mode of communication is required not only for a day-to-day living (basic survival) but also to earn one's livelihood. Writing and reading skills develop much later compared to speech. It is to be appreciated that even illiterate persons can use the speech modality for communication. Unlike illiteracy, the reading skill enlarges one's vocabulary and it can pave way for finer aspects or nuances related to knowledge, emotion and so forth, leading to one's expansion of self-expression via speech.

Broadly, there are five phases for dealing with communication disability: (a) creating awareness on preventive measures, (b) early screening and diagnosis, (c) intervention or treatment, (d) speech therapy or prescription of assistive devices and (e) education and rehabilitation.

6.2.2 Communication disabilities in children and adults

Speech is a highly evolved skillful activity learnt since early childhood. This learning process in early childhood appears to be almost subconscious. Speech modality is often taken for granted because of the ease with which we can communicate. However, not everyone is fortunate to have this skill. Due to congenital factors, there may be a deficiency in the cognitive abilities to learn a language or a disability to comprehend speech or a disability to issue neuromuscular commands to voice and articulate in order to express the basic needs. Even if a person is born with a normal faculty to comprehend and speak, certain developmental (cognitive) disabilities may come in the way of a full-fledged communication development. Inability to communicate leads to frustration and depression leading to behavioral problems, which may be misinterpreted by others as a mental problem. The difficulties faced by a person with a speaking disability may be compared with the difficulties faced by a person visiting a locality where they speak an unfamiliar language or the difficulties while learning a second language. In India, according to 2002 statistics, 2% of the population of children is born with some type of communication disability; 2% of the elderly population suffers from a stroke of which 10% have loss of speech. We will now briefly cover some of the disabilities.

Hearing disability may arise due to one or more causes such as bad habits (smoking and alcoholic consumption) or consumption of unprescribed medication by the expectant mother, hereditary factors, infections, exposure to very loud sounds, physical injury (resulting from insertion of a sharp object such as a pencil into the ear) and finally due to aging. Hearing disability is broadly categorized as "conductive" or "sensorineural." Conductive hearing loss occurs when the sound is not transferred to the inner ear efficiently. Sensorineural loss arises when the sound sensations are not propagated properly to the brain. Sensorineural loss is common in old age due to the degeneration of nerve cells. Auditory neuropathy is a central auditory disorder where the hearing is proper but comprehension of speech is poor. Since a hearing-disabled person has normal speech production apparatus, it is possible for such a person to be able to speak, provided the onset of the hearing disability has occurred after acquisition of language, at least partially. For conductive hearing loss, hearing aids may be used to enhance the audibility. In case of dysfunction of cochlea, a surgical procedure is used to place cochlear implants to convert sound vibrations to electrically stimulate the nervous path. In the child population, however, merely providing a hearing aid or a cochlear implant (as the case may be) is found to be inadequate. The child has to undergo speech–language therapy and be stimulated with a lot of spoken material and taught to react or respond appropriately.

Cleft lip and palate is a condition where a child has a cut in the lip or a hole in the roof of the mouth (palate). Such children have difficulty in swallowing and speaking. The main cause appears to be the effect of poor diet, improper medication and infections present in the expectant mother. A relatively simple orofacial surgery can correct the deformity. It is preferable that the surgery be performed at an early age. If not, the child would have acquired improper pronunciation habits to compensate for the deformity and these habits often remain even after the surgery resulting in improper speech. The child would then have to unlearn the incorrect habits and then learn proper pronunciation by means of speech therapy to restore normal speech, which takes a long time and quite a bit of effort.

Autism is a disorder where it is believed that the nervous paths reaching the cognitive regions of the brain are mixed-up (hardware wiring problem). For example, visual stimuli may give rise to an auditory impression and vice versa. Similarly, there may be a mix-up of motor and sensory regions. The symptoms of autism are repetitive behavior, inappropriate response to sensory stimuli, clinging to the same toy, dress or person, lack of interest to make friends and others. Early intervention of speech and language therapy is essential in addition to other integrated developmental activities and socialization. Poor nutrition of the expectant mother and the baby has been identified as a possible cause for a child to be born with autism.

It is the author's supposition that children with autism pay attention to the minutest detail, both relevant and irrelevant; they are unable to pay just enough selective attention on only the relevant aspect under a given situation. For example, imagine a classroom whose walls contain colorful pictures of animals, flowers, plants and so forth, a teacher wearing an intricate print dress, classmates wearing different types of dress and external noises coming from outside the classroom. In such a situation a child, who lacks selective attention and who pays attention to the minutest detail, is bombarded with overwhelming amount of information within its brain. If this supposition is true, then one possible way to deal with the situation is to stimulate predominantly only one sensory modality at a time while learning and then gradually increase the complexity of the learning environment.

Attention-deficit disorder, hyperactivity, Down syndrome, spasticity, mental retardation and so forth are other impairments leading to communication disabilities.

6.2.3 Communication disabilities in adults

A very common problem with geriatric population is the hearing loss, especially, the high-frequency hearing loss. Due to an exposure to high noise levels and aging, nerve cells (inner ear hair cells) become degenerated. Frequency-selective amplification hearing aids are available. However, there is reluctance to use the hearing aids by the elderly due to the need for frequently changing the cell and a demand for careful handling of the miniature device with flimsy wiring. In addition, the quality of the auditory sensation after a hearing aid fitting is far from what they were experiencing before the advent of hearing loss. Though there may be only a monoaural hearing loss, the sound localization property demands binaural hearing aids.

Elderly persons are also prone to neurological disorders such as ALS or neuromuscular degeneration, Parkinsonism and Alzheimer's, resulting in the loss of memory and speech. There are various objective measures to assess the degree of degeneration that has taken place in which changes in the characteristics of speech are an important indicator. The main causes are supposed to be genetic predisposition, exposure to toxic elements, viruses, infection, stress and so on. Some preventive measures include avoidance of stress, practicing yoga exercises and engaging in progressively challenging and stimulating auditory and cognitive tasks to slow down the degenerative process. (See "Cognitive tests" described in a later section.) However, general neurological disorders (such as motor neuron disease, MND) that affect *all* organs as well as speech are usually irreversible and do lend themselves to preventive measures.

Another common problem is a stroke resulting in a loss of speech production and/or comprehension, generally called aphasia. This may manifest as speaking with gibberish sounds or the use of irrelevant words. Stroke may arise due to a lack of adequate blood supply to the brain by a narrowing of the carotid artery caused by the accumulation of plaque or due to a rupture of thin arteries in the brain caused by uncontrolled hypertension (high blood pressure). A slow and persistent effort is required to rehabilitate an aphasic patient.

6.2.4 Concerns of normal speaking persons

Though we are aware of normal body temperature, normal blood pressure and normal heart rate, are we aware of parameters governing a normal voice? Even more, are there such measures in the first place? Not too often is a person concerned about their voice unless there is a noticeable deviation in voice quality such as pitch breaks and hoarseness. In some cases, an adult male voice sounds like a female (puberphonia) and vice versa (androphonia). This is primarily seen as a functional disorder that can be easily corrected with therapy unless there is an organic problem with the vocal folds. For others, the voice gets tired after prolonged usage. This is common after social functions. However, if this happens by the end of a working day where the voice was not used in any excessive way or if it appears to be a persistent problem, remedial measures should be sought.

Even persons with a normal faculty of speech and language may be desirous of improving their voice quality so as to make their speech more intelligible and more appealing. This is especially true of professionals who use their voice extensively such as actors, teachers, lawyers and singers. A person may not be aware of the full potential of one's own voice. In fact, a good command over language and a good expressive skill gives tremendous confidence to an individual.

6.3 Speech production modeling

6.3.1 Physiology of speech production

When there is a desire to express a thought, the brain formulates sentences in the target language with an appropriate choice of words depending on the social context. Speaking is not merely vocalization of the sentences so formed. Speaking is accompanied with gestures, emotions and eye contact with the audience (see Figure 6.1). The vocal apparatus at the peripheral level consists of the respiratory, the laryngeal and the articulatory systems. The tongue, the lower jaw, lips and velum are referred to as articulators. The brain sends neuromuscular commands to the muscles of vocal apparatus. Much of this activity takes place subconsciously. Even a speaker becomes aware of the words after he/she listens to his/her own speech. There are about 67 muscles associated with speech production. There has to be a coordinated movement of these muscles with time synchrony of the order of a fraction (one-thousandths) of a second. It is like an orchestra of 67 different instruments being played together. Any lack of coordination or out-of-step timing results in a deviant speech.

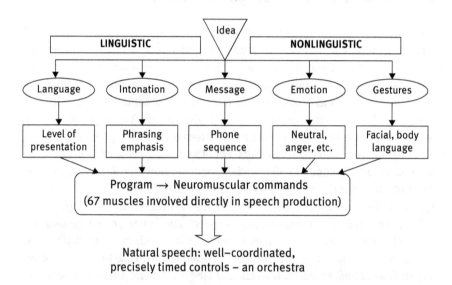

Figure 6.1: Speech communication is a complex process. Speaking is not merely vocalization but involves allied aspects.

The respiratory, laryngeal and articulatory systems are primarily meant for survival. Respiratory system is meant for breathing. A pair of vocal folds located in the larynx acts as a valve preventing food and liquids from entering the lungs. The tongue is used during eating. Nevertheless, these three systems have been cleverly adapted by humans for speaking and singing.

6.3.1.1 The role of respiratory system

In normal breathing, inspiration is supposed to be voluntary and expiration is rapid and involuntary like blowing a balloon and letting-off the air. During speech production, inspiration is rapid but expiration is a voluntarily controlled prolonged act (by a controlled inhibition of the expiratory muscles). Thus, the air from the lungs is slowly released. It is the expiratory air that is modified by the larynx and the articulators to produce speech sounds. If the respiratory control is poor, one has to take frequent breaths while speaking leading to dryness of the vocal folds. The impact of dry folds against each other may damage them in the long run. Also, well controlled respiration leads to a loud and clear voice with an ability to speak/sing for a longer duration without getting tired. Lung pressure is the primary determinant of the level of speech (loud/soft). The volume of air stored in the lungs that can be efficiently utilized determines the duration of speech that can be spoken in a single breath. Abdominal breathing is recommended for speaking and singing. In this type of breathing, during inhalation the walls of the lower abdomen expand and during exhalation the walls recede slowly.

6.3.1.2 The role of laryngeal system

A pair of muscles called vocal folds (vocalis muscle or vocal cords) in the larynx seals the trachea (the passage to lungs) preventing food and liquids from entering the trachea. The same pair of vocal folds in the larynx acts as an important instrument in voice production. During speech production, the vocal folds are set into self-oscillations by a process called "phonation." This is also referred to as "voicing." The process of self-oscillations can be understood easily with an analogy. Blow a balloon and close the lips, save for a narrow opening. Air gets released slowly through the narrow opening. The escaping pressurized air pushes the lips side-ways. In the case of phonation, after a deep inhalation, the two vocal folds, which are normally relaxed and well separated for breathing, are forcefully brought together (tensed) by an elastic force to seal the trachea (medial compression). The excess pressure below the vocal folds, called the lung pressure or sub-glottal pressure, strips open the folds side-ways, and the air escapes through the narrow orifice formed between the vocal folds. The narrow orifice between the vocal folds is called the glottis. As the air gushes out through the glottis, the intensity of the force that separated the folds drops considerably. As a result, the sideway movement of the vocal folds is slowed down and ultimately stops. However, the initial elastic force (tension in the folds that was applied to bring the folds together) now moves the folds inwards and closes the glottis (like the action of a spring door that closes automatically after pushing it open). After a contact of vocal folds, due to inertia, vocal folds remain pressed against each other for a while closing the glottis. The closure may be complete or partial along the length of the vocal folds. This forms one cycle (closed glottis to closed glottis). There are three important phases within one cycle: (i) opening phase where the folds move side-ways away from each other (ii) closing phase where the vocal folds move inwards and approach each other and (iii) closed phase where the folds remain in contact. When the glottis is closed, the excess air pressure in the lungs strips open and separates the vocal folds sideways. This cyclic action repeats itself resulting in an almost periodic movement of the vocal folds required for voice production giving rise to the distinct perception of pitch. Once the two forces, viz., the excess lung pressure and the elastic force are set, the vocal folds go into cyclic vibrations on their own. This is unlike the periodic breathing or heart rate where an electrical impulse is sent from the brain for every cycle.

By controlling the tension in the various muscles of the larynx, the pitch (number of cycles per second) can be varied in a desired manner. This can be appreciated especially in the case of singing and while speaking with a good intonation. Also, the relative intervals of the three (opening, closing and closed) phases during a single cycle determines the manner of vibration of vocal folds, which in turn determines the voice quality. The voice quality can vary from a melodic voice to a breathy or pressed voice.

6.3.1.3 Voice disorders

A simple test done by a neurologist is to ask a person to stand erect with closed eyes and outstretched hands. If there is a neurological deficiency, the hands will tremble, sometimes very vigorously. Similarly, a neurological deficit to maintain the tension in the vocal folds results in an inability to sustain a steady voice, which results in random pitch variations giving rise to a roughness in voice. In the early stages, the randomness in pitch may not be perceivable. But instrumentation (computer with software) analysis can detect and quantify the degree of randomness. Another form of voice disorder is hoarseness or breathiness in voice which leads to quick loss of air supply from the lungs demanding frequent intake of breaths leading to dryness of vocal folds. This arises due to an inability to completely close the glottis during vocal fold vibrations. The opposite of breathiness is the pressed voice. Too much adduction or tension results in vocal folds impinging on each other with a great force resulting in edema or nodule formation of the vocal folds (see Section 6.7.1 described in a later section). Smoking and chewing of tobacco products results in damage to both the lungs and the vocal folds and it may lead to cancer and subsequent removal of the larynx and loss of voice.

6.3.1.4 Role of articulatory system

How are the different speech sounds produced when air from the lungs is merely interrupted periodically by the vibrating vocal folds? The shape of the open space within the mouth (hollow cavity), called the vocal tract (VT), is altered by moving the articulators: the tongue (moving up or down, back or front by various degrees, curling, tongue tip movement), the lower jaw (up/down movement), the lips (spread, protruded) and others. The passage for air within the VT at any desired place (from the glottis to the lips) can be made relatively wide open (for vowels) or narrow or obstructed (for consonants) by appropriately positioning the articulators. The place of narrowing or contact can also be controlled. As the shape of the VT changes, the corresponding sound pattern that emerges also changes. Thus, different speech sounds are produced. In addition, the velum can be used to seal the passage of air to the nasal tract to produce vowels and fricatives or can be opened partially to produce nasalized vowels or opened fully to produce nasal sounds such as "m" and "n."

Merely moving the articulators doesn't produce audible sounds. Exhaled air from lungs (as in breathing) has a very low frequency below the audibility. By means of modifying the exhaled airflow from lungs in various ways, acoustic energy of audible frequencies is generated. There are three distinct manners in which such a modification takes place. Mixed manners are also possible. First, the manner called "voicing" or "phonation" described earlier is used to produce vowels, semivowels ("y," "w"), laterals ("l," "r") and nasals ("m," "n"). In the second manner of airflow modification, vocal folds are kept wide apart, a steady high-velocity airflow is directed against an obstruction or made to pass through a very narrow opening within the VT resulting in a noise like sound. This manner is used for producing sounds referred to as fricatives (e.g., "s," "sh," "ch" and "f"). For some sounds, the air passage at a chosen place within the VT (viz., velar, alveolar, retroflex, dental, labial) is completely sealed; the air pressure is built-up behind the sealing; then the air pressure is released suddenly resulting in an explosion of air (like bursting of a balloon). This manner is used in producing sounds referred to as stops and affricates (e.g., "k," "ch," "t," "th," "p" or "g," "dh," "d," "th" and "b").

6.3.1.5 Speech disorders

In speech production, the articulators move from one set of target positions to another continually. Speech sounds are produced as overlapping gestures and not one sound after another in isolation (i.e., not telegraphically). The inability to move the articulators to the desired positions (targets) or a sluggish movement of the articulators or an inability to produce the desired manner of sound generation results in a deviant speech. The importance of the coordination of the movement of the various muscles as well as the interior structure of the VT can now be appreciated. An extreme case is that of a cleft palate where at birth the tongue gets stuck to the upper palate deforming the oral cavity and causing immobility of the tongue. With such a deformity, speech sounds are naturally distorted. Dysfunction of velum may result in airflow through the nasal tract for all sounds, thus producing a distinct nasal voice. If this is due to bad habit, it may be corrected. But if it is due to some defect in controlling the velar movement, then it may call for surgery. Due to old age, the muscular movements become sluggish resulting in a slow movement of muscles and inability to reach the desired positions resulting in a slurred speech. Another example is vocal palsy (spasticity of muscles) resulting in immobility or slow movement of muscles resulting in speech distortion. In the degenerative neurological disorder (MND or *ALS*), the ability to control the movement gradually deteriorates so much so that even breathing and swallowing become difficult let alone the finer act of the production of speech.

6.4 Physics of speech

6.4.1 Concept of a sound wave and spectrum

We have seen above how the physiological mechanism is used to produce different speech sounds. Speech sounds spread out in the atmosphere as "sound waves" (acoustic waves) to reach a listener or a recording device such as a microphone. The term "acoustic" denotes a form of energy. Speech sounds are acoustic pressure waves of audible frequencies.

There are two distinct (time and frequency) points of view to describe the "complex nature of the waves" associated with a given sound. In the time domain point of view, the instant-to-instant variation in the pressure level is described. This instant-to-instant variation in the pressure wave is called a (continuous) speech signal. A graphic representation of the acoustic pressure of a sound wave (relative to the atmospheric pressure) versus time interval is called "waveform."

Another point of view is called the "frequency domain" point of view. An analogy to the frequency of sound is the color in vision. Sun's white light is a fusion of various colors of varying degrees of brightness. A raindrop decomposes the white light into its component colors resulting in a "spectrum of colors" as seen in a rainbow. The simplest possible sound wave, called as a "pure tone" (like a single color), consists of a single frequency component. Theoretically, a "sine-wave" is a mathematical description of a "pure tone." A complex wave is a bundle of such simple components ("pure tones") of different frequencies. Different sounds have relatively different proportions of energy associated with the component frequencies. The distribution of the energy at different frequencies is referred to as "spectrum." A graphic representation of energy (usually in log scale) versus frequency is referred to as (log) "spectrum."

Speech signal is a sequence of sounds. During speech production, articulators are continually moving. This means that the spectrum of a speech signal is continually changing. For a meaningful interpretation of the spectrum of a speech signal, a sequence of (short-time) spectra of short intervals (20-40 ms) of a speech signal is computed and displayed. A graphic representation of such a sequence of spectra is called "spectrogram."

Our visual perception can be "cheated" by mixing only three primary colors in various proportions as done in computer monitors and TV sets. In the case of a sound, such very few "primary sound waves" are not present. A large number of frequency components of different proportions have to be used to represent speech sounds.

6.4.2 Speech production modeling

A model is a mathematical representation of a physical process. The acoustic theory of speech production attempts to establish a mathematical relationship between the physiological states of speech production mechanism and the acoustic pressure wave of a speech signal (in turn the spectrum of a speech wave). For example, given the position of the tongue, the jaw and the lips during the production of a vowel, the acoustic theory can predict the spectrum of a vowel sound. The classical acoustic theory leads to the so-called source–filter model which considers speech production to be mathematically two independent processes: (i) source process and (ii) VT filtering process.

In fact, the development of the acoustic theory and source–filter model is so well advanced that it is possible to "artificially synthesize speech sounds."

Such a synthesis has two broad approaches. In one approach, called "articulatory synthesis," the human speech production apparatus and its dynamics (the movement) can be simulated (or mimicked) on a computer and the desired speech sounds can be synthesized. In the second approach, called "parametric approach," the pressure signal of a desired sound can be synthesized (constructed) by mixing the different frequency components of the source and VT appropriately.

6.4.3 Parametric representation of speech

The concept of "parametric representation" can be explained with an analogy to human anatomy and the design of ready-made garments. The height, girth and shape of different persons are widely different based on the anatomy. In fact, the number of shapes and sizes of persons in the world is almost innumerable. How to design trousers for all the persons? Broadly, for the purpose of designing the trousers, all persons are categorized based only on the (i) "height from the waist to the feet" and (ii) "waist size." Now the different anatomical shapes of humans are characterized by only two numbers – the height, H, and the waist W. We say that the anatomy of persons is modeled by two parameters, H and W, for this restricted purpose.

Also, an additional approximation is made. Two trousers are designed to differ in height or waist only by 1 cm at a time and very short or very long H and very small or vary large W are not manufactured at all. To use the technical jargon we say the height and waist are *quantized* in steps of 1 cm. The concept of quantization is used in the digital representation of a speech signal to be described subsequently.

The number of speech sounds of different languages as spoken by speakers of different gender and age and individuality is highly complex and the variation is innumerable. In order to be practical, the signal or the spectrum of a given sound is characterized by a certain finite number of parameters. The complex speech wave is described by means of these parameters. The names of these parameters are determined by the mathematical equations used to model complex speech signal as well as to the underlying physiology. There are a large number of such parameters. Some of these parameters are introduced in a subsequent section with specific reference to their clinical significance. Estimating the values of these parameters given a speech signal is called speech analysis.

A complete understanding of the relationship between speech production and the parameters, and inversely mapping the parameters to states of the speech production apparatus, the adequacy of these parameters to capture the complex process of speech production of all languages and of all voice types for all speakers, both normal and pathological, and an accurate analysis to estimate the values of these parameters are not yet fully understood or solved. This is similar to the field of medicine, where research is continually discovering new results relating to the complex processes in the human body vis-à-vis the state of health.

6.4.4 Goal of speech production: articulatory or auditory?

"We speak in order to be heard and need to be heard in order to be understood" goes a famous saying by Jakobson and Waugh. In order to teach a child or a person with a speech impairment, we must know the goal to be achieved in speech production by the speaker. Is "achieving precise articulatory positions" – the goal OR is "producing an appropriate auditory pattern to help a listener decode speech" – the goal? When we learn to speak, we are taught the correct articulatory positions to be used for proper pronunciation of a speech sound. Literature on linguistics (articulatory-based distinctive features) and phonetics (manner and place features) emphasize the importance of speech production. These seem to suggest that "achieving precise articulatory positions" to be the goal of speech production.

On the other hand, some findings, as outlined here, seem to suggest that a speaker's goal is to produce an "appropriate auditory pattern": (i) It is well known that alternate pronunciations (compensatory articulation) may be used for producing one and the same speech sound. This enables one to speak while eating or having a candy in the mouth. (ii) Consider the production of the syllables, "kee" and "koo." Observe the lip shapes when a speaker is about to utter these syllables. The lip shapes formed even before speaking are different for these two syllables. The lip shape when producing the consonant "k" is that of the following vowel. This is called "anticipatory co-articulation." Despite different articulatory positions used for "k" sound, a listener has no difficulty in recognizing the consonant. (iii) Speakers belonging to the same language and dialect are asked to utter one and the same vowel. The x-ray images taken during such a production reveal that speakers adopt different articulatory positions for one and the same vowel. However, listeners correctly identify the vowel. (iv) Similarly, 22 different articulatory configurations have been reported for the consonant "r" based on x-ray images of different speakers.

Different pronunciations produce different acoustic signals. The characteristics of a speech signal for the same speech sound are influenced by the context in which it occurs. Different repetitions of the same sound, in the same context, by the same speaker, also result in differing acoustic signal. Significant acoustic variability arises due to differences in gender and age of the speakers. How then a listener is able to abstract one and the same phoneme despite wide acoustic variability? This implies that the goal of production is not in producing an invariant auditory pattern. We are yet to understand the underlying perceptual phenomenon fully.

One of the ways of dealing with wide acoustic variability is referred to as "normalization." Consider for example visual perception. When a person is walking toward an observer, the height of the person's (inverted) image on the retina increases. However, the person is not perceived as a dwarf when he/she is at a far-off distance or as a very tall person when he/she is close by. Visual perception "normalizes" for the image on the retina for the distance. Similarly, in speech perception, the variability in the spectral pattern is supposed to be normalized for differences arising due to gender and age of a speaker.

Another approach is to look for acoustic correlates (other than the parameters of a speech production model and other than the spectrum) that are context and speaker independent in order to capture the invariant perceptual attributes of a speech sound. This approach has been used in the pronunciation therapy to be described in a later section.

6.4.5 Auditory processing of speech

Speech sounds are received by the hearing mechanism and processed by the brain of a listener. It is generally known that at the peripheral level the hearing mechanism converts sound energy into mechanical movement of a (basilar) membrane within the cochlea, which in turn generates tiny electrical impulses that propagate along a large number of nerve fibers to a specific region (auditory cortex) of the brain. It is also known that there are dedicated nerve fibers for different frequencies and thus the hearing mechanism separates the stream of sounds continually into different frequency components as and when they strike the ear. These electrical impulses then travel to the brain producing the sensation of sound, and the brain (or the mind) interprets these sensations as speech and assigns a meaning giving rise to perception. Since the different frequencies are resolved, the hearing mechanism is said to perform frequency analysis or spectral analysis. There are two major areas as related to auditory processing: (i) psychoacoustics and (ii) perception of speech. There is considerable research advancement in the area of psychoacoustics. However, we are yet to fully understand the phenomenon of speech perception.

6.4.6 Allied aspects of speech production

Apart from the "message" (which has a written counterpart), a speech signal also contains a speaker's voice quality, emotions, dialect and so on. Speech signal processing attempts to capture such allied aspects as well.

6.5 Some important parameters and their clinical significance

6.5.1 Vital capacity and maximum phonation duration or time (MPD or MPT)

The air stored in the lungs is the primary source for producing speech. It is this air which is exhaled slowly and interrupted by vibrating vocal folds and modified by VT during the production of speech sounds. The quantity of air that can be stored in the lungs is called the vital capacity, part of which is utilized during speech production. An indirect measure of the vital capacity is the maximum duration for which one can sustainably say a vowel in a single breath. A low MPD implies that either the vital capacity is low or that the air is inefficiently utilized due to a breathy voice.

6.5.2 Subglottal pressure

In addition to the volume of air, the excess pressure in the lungs, called the subglottal or lung pressure, is another important parameter. Since this pressure is very low, it is expressed in a unit called "cm of water" unlike the atmospheric pressure that is expressed in "cm of mercury." A very low subglottal pressure (less than 3 cm of water) can't bring about the vibration of vocal folds. A moderate subglottal pressure (between 3 and 6 cm of water) results in a weak voice that is hardly audible. For a good clear and loud voice the subglottal pressure must be high (above 10 cm of water). A commander in a field or a street hawker may use about 20 cm of water to produce a very loud voice.

The pressure must be maintained steady during speech production or allowed to decline gradually. A rapidly falling lung pressure results in a lower MPD. An unsteady decline of pressure produces a wobbling or a fluctuation voice.

6.5.3 Speaking level, intensity and loudness

Some voices are loud and clear, and some voices are weak and muffled. This depends on the "level" of speaking. Since the instantaneous amplitude of a speech signal is continually varying and given that different sounds have different complex wave shapes, instead of measuring the level, a parameter called "intensity" (the sound energy of a short interval of a speech signal) is used. Intensity is the amount of acoustic energy that traverses a unit capture area per second. Loudness is a perceptual attribute related to the intensity of a sound wave. A change in the peak level (peak amplitude) of a sound by a factor of 10 is heard as if the loudness has only been doubled. In other words, the loudness is logarithmically related to intensity. Intensity is expressed in a logarithmic scale called sound pressure level (SPL), whose unit is referred to as decibel (dB). Loudness also depends on the frequency of a sound wave. A low-frequency sound (about ten units in amplitude) and a high-frequency sound (about one unit in amplitude) are both perceived to be of about the same degree of loudness. A more precise relationship between intensity and loudness is obtained based on psychoacoustic experiments.

The perceived quality of a loud sound is not the same as an amplified weak sound since the physiological effort used for producing a loud sound is different from that of a feeble sound. Physiologically, the most important parameter that controls the SPL is the subglottal or lung pressure. For a steady vowel spoken at a steady pitch, the subglottal pressure and its steadiness and so forth can be indirectly inferred by measuring the intensity. A change in the physiological effort also changes the manner of vibration of vocal folds.

Intrinsically, different speech sounds have different levels though a speaker does not intend to produce such changes in the level of different sounds. Thus, a fricative "s" may be several times lower in amplitude compared to a vowel sound (e.g., in the word "say"). The peak amplitudes of different vowels are also different. This is because the level of a speech signal is also determined by the VT filtering action in addition to the subglottal pressure. Perceptually, however, we don't perceive such differences. Similarly, the duration of various speech sounds is different. However, we perceive different speech sounds to be the same duration (except for the long and short vowels).

6.5.4 Average pitch, intonation, jitter

An analogy to heart rate may be given in this context. The interval between two corresponding peaks is called the "period." The number of peaks per minute gives the heart rate. The average heart rate is said to be 72 beats per minute. In actual practice, the heart rate varies depending on the physiological need and emotional state. The heart rate is controlled by sympathetic nervous system, and an individual does not normally have the ability to control the heart rate.

As mentioned earlier, during voicing, a pair of vocal folds in the larynx execute cyclic or periodic movement thereby opening and closing the glottis. The rate of vibration of the vocal folds (the number of cycles per *second*, denoted by the unit Hertz, Hz) is called the fundamental frequency and is denoted by the notation $F0$. On an average, $F0$ is about 120 Hz for an adult male, 180 Hz for an adult female and 300 Hz for a child. Such differences arise because of differences in mass and length of the vocal folds.

The rate of vibration gives rise to the "perceptual attribute" called "pitch," whereby a voice is described as bass (low pitch) or shrill (high pitch). The rate of vibration of vocal folds can be voluntarily varied. For example, during singing of the ascending notes, one increases the pitch in steps. During singing of the descending notes, one decreases the pitch in steps. Sometimes an intentional and well-controlled modulation (vibrato / *gamaka*) may be introduced in the pitch.

During speech production, $F0$ is varied over an utterance in a rule-based manner and this systematic variation is called intonation. Forming phrase groups such as the main clause and subordinate clause are marked by the intonation pattern. An assertive sentence has a falling pitch, whereas an interrogative sentence ends with an increased pitch toward the ending of the sentence. Two words constituted of the same phonemes may differ in the meaning because of a change in "stress" arising mainly due to pitch change. For example, "judge" as a noun (The judge sat on the bench) versus "judge" as a verb (don't judge me). This is a semantic stress. When a specific word is emphasized, then also the pitch increases over the main syllable of the emphasized word. Pitch change plays a significant role during an expression of emotion.

Apart from the linguistic, semantic, emotional carrier, pitch is also a very important parameter used for diagnosis of voice disorders. When a client is asked to say a steady vowel, if there are large random variations in pitch, it is indicative of a neurological disorder. A large standard deviation in the pitch of a steady vowel indicates some underlying dysfunction. Jitter is a parameter that quantifies the fluctuations in the pitch on a cycle-to-cycle basis. There are other objective measures to identify the long-time variability on the basis of 3 or 5 or 11 cycles. However, there may be an intentional structured variation (modulation) in pitch. There are methods to distinguish a voluntary controlled structured variation as against an uncontrolled random variation in pitch.

A complete loss of periodicity as in a hoarse voice is called dysphonia or aphonia. Pitch breaks during vocal exercises is an indication of inability to exercise muscular control. Pitch is a robust parameter in the sense that the periodicity can be measured reliably even when the speech signal is distorted during the recording or when a speech signal is recorded over a telephone.

Occasionally, the level of pitch used by a person may not be appropriate for the age and gender. For example, a male speaker may sound like a female. As mentioned earlier, this is called puberphonia. That is the male speaker retains the prepuberty voice. Conversely, for a female voice androphonia is when the female speaker sounds like a man due to a very low pitched voice. Another empirical concept is the "optimum pitch." It is empirically hypothesized that every individual has an "optimum pitch level" depending on the anatomical size of the larynx and the VT. If a person uses such an optimum pitch level, then he/she can speak for longer duration without the voice getting tired or strained, speech has a greater intelligibility and clarity. One should try to use the optimum pitch level. Infant cry analysis, especially for the pitch level and its range, is a precursor to predict if a child has a hearing disability.

6.5.5 Morphology of voice source and glottal parameters

The term morphology refers, in general, to a "pattern" in a signal. For example, given the electrocardiograph (ECG), a cardiologist visually examines the graph to see if the "pattern of ECG" is normal or deviant. If deviant, the cardiologist is also able to infer the functionality merely by looking at the "pattern of ECG." This is called morphological analysis. Similarly, a signal corresponding to the manner of vibration of vocal folds can be computed given a vowel signal using a technique called "inverse filtering." By visual examination of the pattern of the voice source signal (morphological analysis) one can infer the status of the vocal folds (e.g., presence of an edema or a nodule). Further, the relative intervals of the three phases (temporal glottal parameters) help to characterize the vocal fold vibratory pattern to detect different voice qualities such as breathy voice, pressed voice or a falsetto voice as already noted.

Morphological analysis relating to voice is also done using an instrument called electroglottography. Here, two electrodes are externally placed across the neck nearly at the level of the vocal folds. A low-intensity high-frequency electrical signal is passed through the electrodes. When the vocal folds are in contact, there is a low impedance path for the electrical current, whereas when the vocal folds are apart the impedance increases. As the vocal folds vibrate, a modulation is produced in the high-frequency signal. The modulating component, referred to electroglottographic (or EGG) signal gives an indication of vocal fold contact area. Various organic voice pathologies can be inferred by a morphological analysis of the EGG signal.

6.5.6 Spectral parameters

We have already noted that a complex wave of speech has innumerable number of frequency components. The energy at different frequencies constitutes the spectrum of a speech sound. Some sounds can be pronounced in a steady manner like vowels, laterals, nasals and fricatives. The spectrum in the steady part of the sound may be considered to be the representative spectrum for that sound.

The spectral shape consists of two components: (i) a gross spectral shape or spectral envelope and (ii) fine spectral components. Vowels and voiced consonants show clearly defined peaks in the spectral envelope. For fricative and stop sounds, spectral envelope is diffused. The significant peaks in the spectral envelope that arise due to the resonances of the VT are referred to as formants. The locations of the formants depend on the sound being spoken. Also, the locations of formants are influenced by the gender and age (anatomical size) of the speaker. There are about four significant formants for vowels within 4000 Hz, of which the first two formants characterize vowel identity.

The fine structure of the spectrum of a voiced sound shows regularly spaced peaks–valleys. The peak locations are referred to as harmonics. For a periodic signal with a pitch $F0$, these harmonics are located nearly at $F0$, $2F0$, $3F0$ and so on. For a good resonance voice, one of the harmonics coincides with the formant frequency (pitch tuned to a resonance) thus boosting the energy level making the voice sound "bright." For a steady voice, harmonics are regularly spaced up to a very high frequency. However, for an unsteady voice, only a first few harmonics are regularly spaced. Regular versus irregular harmonic distribution is captured by parameters such as "harmonic-to-noise ratio" (HNR), "glottal-to-noise excitation ratio" (GNE) and dysphonia index (DSI). The relative difference in spectral levels at the pitch and the formants is another indication of the voice quality. The perceived quality of voice depends on the spectral shape in addition to temporal parameters already covered. Thus, for example, a breathy voice has a very low level at high frequencies. This is characterized by the spectral slope parameter. Some singers are able to increase the spectral level in the frequency range of 3000 to 5000 Hz. Such a boost in the energy at high frequencies is referred to as "singer's formant." Because of the sensitivity of auditory system to high frequencies, such voices are perceived to be loud.

6.5.7 Normative data

Consider as an example "blood test." Certain clinical parameters like blood sugar level, WBC, RBC (blood count), creatinine and cholesterol levels are measured.

By "the analysis of blood samples" on a large number of clients with normal health, the normal range of these parameters is determined to constitute normative data. By the analysis of blood samples of clients with pathology, a relationship is established between the range of a parameter and the type of pathology. In a similar manner, analysis of a speech signal estimates or measures the parameters of a speech production model. By recording "normal speech" or "normal voice" samples and analyzing them one arrives at the "normal range" of values for these parameters. By analyzing the "deviant speech" or "deviant voice," a relation between a specific type of deviation and the range of the values of these parameters is derived. There has been a tremendous success in terms of establishing the normative data with respect to various parameters.

An alternative to the above type of objective analysis is the use of a questioner. Here a series of questions are posed to the client and a marking, in the range of 0–5 (or 0–10), is assigned by an experienced speech language pathologist. Subsequently, the scoring is analyzed to arrive at the underlying pathology. Different schools use different sets of questions and scales.

6.6 Digital speech signal processing

6.6.1 Digital representation of a speech signal

The link between a speaker and a listener is the speech signal. In terms of technology, using a microphone, sound waves can be converted into an electrical signal. The strength of an electrical signal (either current or voltage) is continually (instant to instant) varying with respect to time as the acoustic pressure falling on the microphone keeps changing due to the sound waves. The continuous signal is also referred to as an "analog signal." Before the advent of digital technology, measurements were made on an analog signal using bulky instruments. Also, it used to take considerable time to male sophisticated measurements.

With the advent of digital technology, a continuous signal is converted into a discrete set of samples. This is similar to a movie where action seems to appear continuously though only a finite number of "frames of pictures" are shown (30 frames per second or so). In a movie, we are exploiting the so-called persistence of vision phenomenon. However, in digital representation of a speech signal, there is a strong theoretical basis (sampling theorem) and it is claimed that no information is lost though a continuous (band-limited) signal is represented by a finite number of samples. This theoretical basis can be appreciated by the equation governing a curve in geometry. Consider for example a parabolic curve. Although there are infinite number of points on a parabolic curve, information of the locations of any three points is adequate to draw the entire parabola. Similarly, in the case of a speech signal, if our interest is to study a speech signal up to 4000 Hz, then theoretically 8000 samples per second is shown to be adequate. This is called the "sampling theorem."

At each discrete sample, the amplitude of a signal can take on any value in a continuum (innumerable possibilities for specifying amplitude to a fine precision). For practicality, the amplitude of these samples is approximated (or quantized) by integer numbers. Such an integer approximation of the amplitude of discrete samples is referred to as a digital signal. Digital signals can be stored in a computer memory and computations can be carried out on the digital samples using a digital computer. With the digital representation of a speech signal and digital computational power, processing of a speech signal to estimate the parameters has become superfast. See description of "Spectrogram" later for a historical note on analog versus digital technology.

6.6.2 Some signal-processing tools

Signal-processing algorithms are mathematical equations (algorithms) meant for computing spectra or to estimate the parameters of a speech production model or acoustic correlates and others. The implementation of an algorithm is done by software coding in a computer. The software consists of the operational code along with a "user interface." The user interface provides a facility to examine the values of the parameters in a text format or to visualize the parameters graphically or to visualize a speech signal or its spectrum or the dynamically varying spectrum. Thus, software is developed to address the various issues relating to voice and speech production and perception attributes both for normal and deviant cases. We present an overview of some selected signal-processing tools.

6.6.2.1 Spectrograph

Spectrograph is an instrument that converts a speech signal into a "visual pattern," called a spectrogram. The development of a spectrograph instrument was motivated by a desire to help the hearing impaired. It was thought that a hard of hearing person can be taught to "read the visual pattern of speech," thereby decode the message as in writing. However, due to a wide acoustic variability and continual movement of articulators, the visual pattern is extremely complex. Reading a spectrogram to infer the message is an extremely challenging task and only a few experts in the world are able to do it. However, spectrograms were used in early speech research to understand the acoustic to phonetic relationships. These inferences remain valid still today.

Spectrograms were generated using a bulky instrument, printed on a special thermal paper. It used to take about 10 min to print out a spectrogram of speech of 1 or 2 s in duration. With the advent of digital signal processing techniques like fast Fourier transform and digital computers, a spectrogram can now be generated, as a speaker is uttering a sentence, that is, in real time, to be visualized on a computer monitor. Spectrograms are still being used as a visual representation to make broad deductions. A spectrogram can show if speech sounds are well timed or produced with clarity. Also, one of the modes of the spectrogram, called narrowband spectrogram, can show the pitch and its variations to deduce if a vowel has a steady voice or random fluctuations or systematic modulation. Also, intonation pattern of an utterance can be inferred. Spectrograms may be utilized for therapy, provided the user is trained to interpret the visual pattern.

6.6.3 Speech analysis and inverse filtering

Estimating the parameters of a speech production model is called speech analysis. There is abundant amount of published research papers and books on speech analysis algorithms. These techniques are also highly mathematical. We only mention a few of the most popular algorithms: (i) pitch estimation using autocorrelation function, (ii) pitch estimation using cepstrum (read as kep-strum), (iii) estimation of spectral envelope (gross spectral shape) and consequently the formant data using cepstrum and linear prediction (LP) techniques and (iv) inverse filtering technique for deriving voice source signal and subsequently estimating glottal parameters. Interested readers may refer to the published literature for details.

6.7 Software solutions for voice and speech disorders

Many software programs are available in the area of voice and speech disorders. We now describe a set of software programs developed by the author. The author is an electrical communication engineer conducting basic research in the area of speech signal processing. For more than two decades, the author has interacted with speech language pathologists, ENT doctors, teachers of the hard of hearing and special educators. With this background, the author has developed a set of software products. The software is being used in hundreds of clinics, institutions, hard of hearing schools, by professionals in the field almost on a daily basis. In addition, the software is being used as a pedagogical tool for training as well as in research projects, master-level dissertations and doctoral theses works. The software implements speech analysis techniques published in the literature with certain modifications so as to serve the specific applications being addressed. Also, some of the techniques implemented are based on an original research conducted by the author. In this chapter, the author presents an overview of the software as applicable to the area of voice and speech diagnosis and therapy. Readers may visit the URL vagmionline.com or voiceandspeechsystems.com for details.

6.7.1 Vagmi voice diagnostics software

By means of voice analysis, one can ascertain the type of voice disorder. Voice disorders are broadly categorized as functional, neurological and organic. It is relatively easy to correct a functional disorder with therapy. A neurological disorder needs further investigation and has to be handled jointly by a neurologist and a speech language pathologist. An organic disorder implies damage to the vocal folds. This has to be further ascertained by "direct examination" of the vocal folds by an ENT doctor or a phono-surgeon and it has to be decided if a surgical intervention is required or voice therapy exercises would help.

In Vagmi voice diagnostics software, there are two modules: (a) voice analysis – here a steady vowel sample is collected from the client. Later, this voice sample is analyzed. A set of 29 parameters are measured. The results are interpreted. (b) Voice measurements – here measurements are made while the client is present at the laboratory. Examples are indirect measurement on the sustenance of lung pressure, maximum phonation duration and phonetogram.

6.7.1.1 Voice analysis module

Voice analysis consists of the following steps: (a) collection of a voice sample, (b) registration of the client, (c) analysis, (d) report generation and examination of the waveform and spectra and (e) interpretation and suggested course of action. These stages are briefly explained as follows.

(a) Collection of a voice sample: a steady vowel "aa" of about 4 to 6 s in duration is recorded as a voice sample for further analysis. An appropriate microphone has to be used if one is interested in interpreting the waveform of voice source signal computed by inverse filtering. If recording is done using software other than Vagmi, option for compression (mp3) must be avoided and linear pulse code modulation (PCM) monosignal of 16 bits precision at a sampling frequency of 16000 Hz is to be used. If recording is done on a laptop, power connection must be removed to avoid line frequency (hum) influencing the recording. Client must be wearing shoes or else must be seated on a wooden chair with the feet not resting on the ground. A quiet room for recoding is to be preferred. Preferably a microphone stand has to be used. If microphone is held in the hand, it must not be moved during recording.

After recording a sample, the steady part of the signal must be selected for analysis. Steady part of vowel "aa" of about 3 s is to be selected avoiding silence intervals, voice build-up, voice decay etc. If a client has difficulty in sustaining a vowel for 3 s, at least 1 s duration of voice sample must be available. The selected part is saved in a file. Appropriate convention must be followed while giving file names.

(b) Registration: Associated with each voice sample file, the details of the client's name, age, gender and so on are saved.

(c) Analysis: The input file name containing the voice sample to be analyzed is specified by the user. There are several stages in the processing of a voice sample: inverse filtering, $F0$ and intensity extraction, jitter and shimmer analysis, long-term average spectrum of phonation and voice source, HNR and GNE computations and others. There are two approaches to analysis: (i) noninteractive – all stages of analysis are performed without a need for an interaction from the user and (ii) interactive – here, after every stage of processing, the intermediate results are displayed. The user can interact with the program and change the settings to verify the accuracy of estimation. If required, the settings can be changed to correct or improve the accuracy. Subsequently, the user moves to the next stage of processing. For example, there are two types of inverse filtering: formant based or LP based. The number of LP coefficients can be specified. Formant data can be edited. The inverse filtered signal and spectrum of phonation signal are displayed. After ascertaining that the appropriate results are obtained, then the user goes to the next stage. Figure 6.2 shows the estimated voice source for the same voice sample using these two types of inverse filtering. In this example, the voice source obtained using LP-based inverse filter is noisy and shows some artifacts, whereas the voice source obtained using the formant-based inverse filter is closer to the expected wave shape based on the physiological process.

Similarly, while extracting $F0$, a set of default variables is used. The extracted $F0$ contour is displayed. Vowel sound is synthesized with the extracted $F0$. This can be compared with the original voice sample to detect any gross errors in the estimation of $F0$. Validation of $F0$ estimation can also be made using a narrow-band spectrogram. If required, the variables like frame length (block duration), pitch range, choice of signal to be processed for extracting $F0$ (voice source or

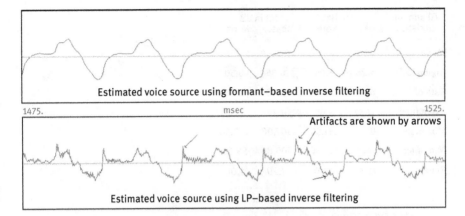

Figure 6.2: Inverse filter output for a natural vowel using formant-based and LP-based methods. Formant-based method gives a signal closer to the physiological process. In some cases, LP-based method gives artifacts.

phonation signal, full-band or low-pass filtered, etc.) can be edited until an accurate estimate of $F0$ is obtained. Similarly, jitter and shimmer parameters may be measured based either on voice source (production based) or phonation (perception based) signal.

(d) A clinical report is generated. The measured parameters are compared against normative data. Figure 6.3 shows such a typical report. The results can also be examined graphically.

(e) Interpretation of the analysis results to arrive at a diagnosis is the responsibility of the examiner. This depends on the clinical experience of the examiner.

6.7.1.2 Voice measurements

The following measurements are made when the client is at the laboratory: (a) lung (or subglottal) pressure sustenance – this is an indirect measurement technique. For a steady vowel with a steady pitch, the measured intensity is proportional to the lung pressure. The duration of phonation, the rate of declination of lung pressure, the standard deviation compared to a linear declination are measured. (b) Maximum phonation duration – here the client is asked to produce a sustained vowel as long as possible. The intensity level must be above a threshold and there has to be periodicity in the signal. The same voice sample is used for estimating the jitter that is required for computing DSI. (c) Phonetogram – here the client is asked to say a vowel at different intensity levels and at different pitch. The range of voice is displayed. Histograms of intensity and pitch are

FO and int statistics	FO in Hz Meas.	FO in Hz Norm.	Int in dB Meas.	Int in dB Norm.
Mean	100.29	128.50	106.48	100.00
Sigma (S.D)	*4.26*	3.00	0.35	0.50
Rate of fluct.	*6.00*	3.00	3.00	5.00
Ext of fluct.	*11.51*	3.00	*1.19*	1.00
Maximum	105.96	131.50	107.08	100.50
Minimum	94.12	125.50	105.88	99.50
Range	*11.84*	6.00	*1.20*	1.00

	dB	Linear
HNR from cepstrum	25.25	18.29
Subharmonic energy	0.08	−22.44

Jitter statistics	Meas.	Norm.	Shimmer statistics	Meas.	Norm.
Mean FO	100.56	128.50	Mean amp dB	115.84	100.00
Jitter T0%	1.06	3.00	Shimmer dB	0.08	0.50
Jitter FO%	1.07	3.00	Shimmer Lin	0.95	1.06
PSigma	*4.34*	3.00	PSigma	0.19	0.50
PVI	*1.89*	0.50	AVI-dB	−0.33	0.05
dpq/dqf%	15.73	50.00	DPQ%	24.72	60.00
DLT	0.08	30.00	DLA	*104.65*	30.00
RAP-3	*0.44*	0.30	APQ-5	*0.60*	0.50
RAP-5	*0.48*	0.10	APQ-11	*1.30*	0.30

Figure 6.3: A typical report generated by Vagmi voice analysis software. Measured values not within the normal range are shown in italics.

displayed. The lowest intensity and the highest pitch are marked by the user. (d) DSI is computed using an empirical formula involving the above parameters.

6.7.2 Vagmi therapy suite

Corrective therapy is for clients who have noticeable problem in the use of voice or in the pronunciation of sounds or stuttering (disfluency) or nasal voice, and so forth.

Enhancement therapy is for clients who sound normal but would like to enhance their voice quality. Also, some clients may believe that their voice is normal but may be abusing their voice inadvertently. It is a common experience of teachers, lawyers and so on that their voice gets tired, even hoarse by evening. Singers would like to comfortably sing for long hours without strain, maintaining their voice quality. Some speakers sound sleepy. Their voice is monotonous. Such problems are addressed in the enhancement therapy module.

Vagmi therapy module is being used in hard of hearing schools, special schools with autistic, MR, down syndrome spastic children and also by normal children having problems. Some general features of the therapy software are briefly presented. A general awareness about the voice and speech production is presented. A set of demo files is available, which can be run for each therapy menu option to know how to achieve the task. Also, online help document is available to learn a task. The values of the parameters currently achieved by the client are measured. These are compared with the normal range of the parameters for the same gender and age group. Goals to be achieved can be set in a graded manner and the level of the task may be set to "easy" to begin with and then set to "difficult." The therapy is made interesting and stimulating by means of games or challenging tasks. The performance is scored. A logbook of the therapy modules chosen and the scores achieved are saved with a file name specific to the client. When a task is achieved successfully a positive reinforcement is shown as a feedback, else a negative reinforcement appears. It has been reported by speech therapists that the use of interesting games and tasks considerably reduces the learning time. Thus, for example, children who used to take several months to learn proper production of fricatives have been able to learn the task within a fortnight. The therapy suite is divided into several modules based on the three physiological subsystems of speech production. A brief description and highlights of these modules is given below.

(i) *Breath support module*: This module is meant for those who take frequent breaths while talking, who are unable to speak for long duration, experience tiredness in voice, experience dryness of vocal folds and for those who have a "breathy or leaky voice." It measures the lung pressure sustenance, maximum phonation duration and efficiency of voice. Four menu options exist for therapy practice for increasing the maximum phonation duration, one menu option to avoid breathiness in voice, two menu options for practicing sustained "s" and "z" sounds. We present an example. Let us say the task is to improve the maximum phonation duration. The game is to complete a 16-piece puzzle. The goal to be achieved is set. As the client says a sustained vowel, parts of the picture are revealed at random locations. When the task is achieved, the complete picture appears on the screen.

(ii) *Voice control module:* This module is meant for those using inappropriate volume and pitch in speech leading to lack of clarity, voice abuse, tiredness of voice, poor control or command over voice. Speech becomes pleasant and effortless with the appropriate voice volume, neither too loud nor too soft and when the voice is not fluctuating. This module consists of several options to control the level of speech and to control the pitch. We present an example. Consider, for example, learning to use the appropriate pitch level. Here the optimum pitch level to be practiced is set as the goal. If required, a tone at the set pitch level may be played over the headphone to tune the voice. When the client says a vowel at the correct pitch, a bird sits on a perch, else the bird flies away from the perch (see Figure 6.4).

(a) (b) (c)

Figure 6.4: Pitch correction. As the client says a steady vowel, the pitch is measured and if it is within certain limits (5% or 10%) of the goal (pitch to be achieved), the bird sits on the perch and score improves. Else the bird flies away from the perch and the score decreases. It is to the left (right) of the perch when pitch is lower (higher) than the goal.

Another unique feature of this module is its usability for individuals whose pitch level is not appropriate to the gender and age (puberphonia: male sounding like a female voice; androphonia: female sounding like a male voice). Here, the client reads a passage and his/her presently used voice is recorded. Then, the recorded voice is scaled down or up in pitch and played back to the client to make the client realize how he/she would sound if appropriate pitch level is used. The client is then asked to imitate the modified voice.

There are two important, almost independent attributes of voice that everyone must be aware of, namely, the level and the pitch. When a person is asked to increase the pitch, they instead speak louder. Or if a person is asked to speaker louder, he/she may increase the pitch. A therapy menu option to control only the pitch or only the level or simultaneously both the pitch and level creates an explicit awareness of these two attributes and also enables one to control them independently. This is especially important for professionals who use voice extensively.

(iii)   *Advanced voice control module*: This module is designed for professional voice users who use voice extensively for their livelihood such as teachers, actors and singers. This is a package to enhance the overall command and quality of one's voice range. It consists of phonetogram and glottogram. In phonetogram, a plot of intensity level in dB versus pitch in Hz is displayed as the client speaks. *Phonetogram* shows if the voice is restricted. Using the visual feedback, the full potential of a client's voice can be explored. Also, it creates an awareness of the two attributes, level and pitch as independent controls. *Glottogram* displays the voice source pattern used by the user, indicating the three phases of a glottal cycle, namely, the opening, the closing and the closed intervals. Typically, these intervals must be almost equal. Also, within the closing phase there are two subintervals indicating if a glottal chink is present or if a breathy voice is being used. Using the graphic illustration as a feedback, a speaker may experiment changing the voice quality till a pattern corresponding to the "modal" type of voice is achieved. Also, this option demonstrates different voice qualities and their corresponding voice source patterns. An appropriate microphone has to be used in order that the wave shape of the inverse filtered signal is interpretable.

(iv)   *Intonation and accent module:* Pitch or $F0$ changes continually while an utterance is being spoken. $F0$ versus time is called as "$F0$ contour." During unvoiced sounds such as "s" or "sh," $F0 = 0$ and hence there will be breaks in $F0$ contour. An imaginary smooth curve over an entire utterance that fits $F0$ contour during voiced parts is referred to as intonation. Accent is determined not only by the intonation pattern but also by local increase in pitch over a syllable, referred to as stress, duration of syllables and so on. This module can be used (a) to visualize $F0$ contour and practice intonation or accent or stress pattern of a spoken utterance, (b) by those whose voice sounds monotonous or sleepy whose $F0$ contour appears almost flat, (c) by hard of hearing clients to make them aware that they can speak with a lively voice and (d) by call center personnel to practice appropriate stress pattern. Using a split window display, a reference pattern can be saved in the upper window. The pattern achieved by the client can be displayed in the lower screen. Using the visual feedback, a client can learn to reproduce the reference pattern. In addition, the same program can be used for displaying Intensity versus time graph to distinguish short versus long vowels. A graph of intensity versus $F0$ can also be generated. A systematic variation is seen in intensity versus $F0$ graph for a good voice.

(v)   *Pronunciation assessment (picture-word articulation test) module:* Picture-word articulation is a standard test to be conducted by a special education teacher, a therapist or a knowledgeable parent. The purpose is to evaluate the correctness of pronunciation of speech sounds of a student or a client. Once an evaluation is done, sounds that are not pronounced properly can be taught and pronunciation corrected using the pronunciation module.

Assuming that each sound has a characteristic spectrum, pronunciation errors can be detected by comparing the spectrum of a reference correct pronunciation against the spectrum of a test sound whose pronunciation is to be verified. Though conceptually, this objective approach appears straightforward, in practice it is not implementable. Two spectra that look almost alike may differ perceptually significantly, whereas two spectra that appear to be distinct may be perceived to be the same speech sound. `At present, the evaluation of correctness of pronunciation is done manually by the supervising person.

By following simple rules, the test can be prepared in any language or for any group of words. This module is also useful for evaluating the speech produced by hard of hearing and post-surgery cleft palate and special children. This module can also be used as a "Phonic Drill" for normal children in schools, for adults learning to pronounce words of a new language and for stroke patients who have lost their speech by helping memory-recall by suppressing the word display and audio clues and eliciting a response. For this purpose, pictures and words belonging to different groups such as parts of the body, daily usage items, animals and fruits can be designed.

(vi) *Pronunciation (vowels, fricatives and plosives) module:* Inability to pronounce certain sounds or substituting one sound for another is not only a problem among the hard of hearing (who rely mostly on lip leading) but also a common problem in case of children with normal speech production and hearing. Even some adults may have difficulty in pronouncing sounds from a language other than that of the mother tongue. Some speakers may not pronounce the vowels distinctly and clearly resulting in slurred speech. Due to bad articulatory habits (clenched teeth, nasalized speech, excessive lip-rounding), speech may lack clarity. Listeners may ask the utterances to be repeated. Some speakers might have difficulty in pronouncing consonants clearly and rapidly due to lack of articulatory agility. With the help of pronunciation therapy, the production of speech sounds is made simple and interesting using tasks, games and appropriate visual feedback.

In the option *articulatory agility* measurement, a client is asked to say a syllable repeatedly, as fast and as clearly as possible. The rate of production of the syllables is measured. Also, a measure called "rhythmicity" indicates if the syllables are produced at a regular timing. By using different syllables, articulatory agility of different parts of the tongue or the movement of lower jaw or the lips can be tested. In "voice focus," a plot of second versus first formant frequency ($F2$ vs $F1$) of vowels is displayed as the client utters different vowels. The extreme positions of vowels can be marked to measure the area of vowel triangle. For a good voice, vowels must be well spread out. For slurred speech or when a bad articulatory habit is used, such as clenched teeth, the corner vowels occupy a restricted space. The area of a vowel triangle has also been used to estimate the degree of neuromuscular degeneration in the case of ALS patients.

There are several options for vowel therapy. We describe one of the options, based on research findings by the author: acoustic correlates for vowels corresponding to front/back high/low dimensions of tongue body movement and lip rounding gesture are estimated. The tongue shape used by the client is superposed against that of reference shape. In the front view, the rounded or spread-out lip shape is shown. Instructions appear on the screen on how to position the articulators for correct pronunciation (see Figure 6.5).

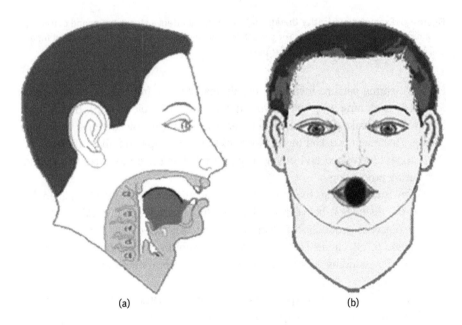

(a)   (b)

Figure 6.5: An example of vowel therapy: (a) As the client utters a vowel, part of the tongue body shape is estimated and shown (in black) against the model tongue shape (in dark gray) for the intended vowel. (b) The front view is shown to indicate lip rounding.

A common problem in children is the mispronunciation of fricative sounds "s," "sh." They may substitute dental fricative "th" (as in "thin") in place of "s." There are several options for fricative therapy. For example, when a client correctly produces a steady "s" sound, a train moves emitting smoke else the train remains stationary (see Figure 6.6a). After learning to say the fricatives, the distinction in production of "s" and "sh" can be taught. When a steady "s" sound is said correctly, a picture of a **s**nake appears, whereas when a steady "sh" is said correctly, a picture of **sh**eep appears (see Figure 6.6b).

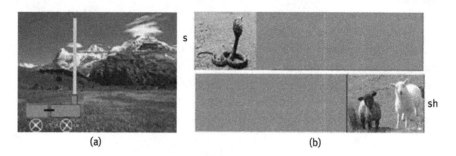

(a) (b)

Figure 6.6: Examples of fricative therapy. (a) As the client utters a steady "s" sound correctly, a train moves emitting smoke. (b) For correct "s" pronunciation, a "snake" appears and for a correct "sh" pronunciation, "*sheep*" appear.

Hard of hearing persons invariably use lip reading. The front shape looks similar for words beginning with "p" (unvoiced stop) and "b" (voiced stop). Hence, they are unable to make a distinction between voiced and unvoiced stops. In plosive therapy, the client is asked to say words like aka, aga, apa and aba. Different patterns appear for voiced and unvoiced plosives helping them to make the distinction during production

There is a hypothesis that mispronunciation arises because the client is unable to hear the distinction between similar sounding consonants. It is said that "perception precedes production." Using a hearing test called "Discrimination test" (see Section 6.7.5), clients can be tested for their ability to hear the differences between similar sounding sounds and also trained to improve their discrimination.

(vii) *Nasality module:* A special two-microphone setup is used to measure nasality. Two microphone capsules are mounted in a wooden handle, about 10 cm apart. A wooden baffle separates the two capsules. The client holds a specially shaped wooden baffle against the upper lip and speaks. The upper capsule predominantly picks up sound waves from the nostrils whereas the lower capsule predominantly picks up the sound waves from the mouth. Thus, the relative levels of nasal and oral outputs are measured. Nasalance is defined as the ratio of the energy output from nostrils to the total energy. For vowels, the nasalance is expected to be less than 10%. For a nasal voice, the measured nasalance will be large for vowels and nonnasal passages. Such voices are called hypernasal voices. On the other hand, nasalance must be higher than 80% for sounds like "m" and "n." In the case of hyponasal voice, the nasalance will be low for nasal sounds like "m" and "n." In the nasalance therapy module, the nasalance can be measured and also an interesting game may be used to control the nasality, provided the velum can be voluntarily moved.

(viii) *Disfluency module:* Some persons are unable to speak fluently though their speech production organs and hearing mechanism is normal. In the case of stuttering or stammering, speech production is disrupted or the rate of speech is nonuniform or a speaker uses fillers like "aa," "mm" or seem to require an extra effort to speak with or without facial grimaces. Such involuntary disruptions, referred to as "blocks of speech," are known by the technical names such as "repetition," "hesitation," "prolongation" and "substitution." For example, a speaker may get stuck on a syllable and keep repeating it before regaining fluency. Or, a speaker may suddenly pause, continue to vocalize (like "mmm") but not say the words (for about half-a-second or so) but later continue to speak. A speaker may take considerable time to initiate voicing or begin the first syllable.

There may be diverse underlying causes for stuttering such as difficulty to initiate voicing or hard glottal attack, tensed vocal folds; disruption of airflow during speaking, tensed articulators, lack of muscular coordination, lack of coordination between speaking and hearing, problems arising due to auditory feedback and psychological disposition.

Several strategies of therapy have been implemented: voice initiation, smooth airflow, prolongation, pause-and-talk, controlled rate of speech, delayed auditory feedback (DAF) and so on. For example, one of the difficulties is in the initiation of voicing. In voice initiation therapy, subject has to begin voicing only after an indicator turns green in color. The green indicator appears at a random instant. By measuring the delay in voicing from the instant the green indicator appears, the degree of difficulty in initiating voicing can be measured. In controlled rate of speaking, a slider moves below the text of a sentence to be spoken at a slow speed. The client has to keep pace with the text just above the slider. The rate of speech, delay in initiation and termination are measured. In Delayed Auditory Feedback therapy, the, client's speech is delayed and played back over headphones (like an echo). The delay for playing the echo can be controlled.

6.7.3 Speech science lab

Speech science lab is a yet another software meant for graphic visualization of speech wave, spectra, spectrograms as well as tools for speech analysis and synthesis. It can be also be used to run an experimental course on phonetics using a set of sample files and a manual. We describe some select experiments on analysis–synthesis in the context of voice and speech disorders.

6.7.4 Analysis–synthesis

As noted earlier, once a description of a speech signal is available in a parametric form, a speech signal can be synthesized. This is called parametric synthesis. Such a synthesis tool can be used for a variety of studies. Speech signal of a "normal voice" can be recorded and analyzed to estimate the parameters. During reconstruction or synthesis, the parameters may be perturbed or manipulated in a controlled manner to synthesize a modified synthetic speech signal. The modified synthesized speech signal can then be heard to arrive at the effect of the perturbation of the parameter chosen on the speech intelligibility or voice quality. Such experiments provide us with insights into the perceptual processes. In an articulatory model for synthesis, it is possible to "mimic" the speech production process. Thus, the position of the tongue, jaw and lips can be set and the rate of movement of these articulators from one configuration to another can also be controlled. Given successive VT configurations, a speech signal can be synthesized. The effect of the articulatory positions, the timing and the rate of movement, on the perceived quality of speech can be studied. We give a few examples on the applications of synthesis tools.

6.7.4.1 Development of voiced/unvoiced distinction of stops in children

One of the topics in perception is how listeners learn to distinguish between voiced stops like [b,d,g] versus unvoiced stops [p,t,k]. One of the parameters that characterizes the difference is called the voice onset time (VOT). Normal production of syllables of "apa," "aka," "aba" "aga" and so on are recorded. These normal recordings are then manipulated to generate a set of stimuli with different values for VOT. The stimuli are randomized and played. Listeners are asked to identify if the stimulus is "p" or "b" ("k" or "g," etc.). Thus, the role of VOT on the discrimination of voiced and unvoiced stops is deduced. Using the adult listeners' response as a reference, the same stimuli are played to young children to find out at what age children are able to match the adult performance. Thus one can infer the developmental age in children at which they can distinguish between such pairs of sounds.

6.7.4.2 Correcting speech of hard of hearing to improve naturalness

Consider another experiment. It is known that a speech signal produced by a hard of hearing person lacks naturalness or fluency. A hard of hearing person's speech is recorded and analyzed. During reconstruction or synthesis, the duration, pitch, formant data are corrected, one at a time, to match with those of normal speech. Thus, the hard of hearing speech is made to sound as nearly close to that of a normal person's speech. By this experiment, one can infer which attribute was lacking in naturalness in a specific hard of hearing person's speech. Therapy is then given to control that particular attribute to provide a significant jump in the naturalness.

6.7.4.3 Subjective judgment of perturbations

Vowel sounds can be synthesized by specifying the parameters such as pitch, intensity, glottal parameters and formant data. During synthesis, random perturbations in one or more parameters can be introduced to synthesize vowels. Such synthesized voice samples may be used to train a specialist to learn to recognize the type and extent of perturbation. Also, the synthesized speech vowels may be used to validate voice analysis software. The synthesized vowel is given as an input to the voice analysis program; The voice analysis program must be able to correctly estimate the type and extent of perturbation used during the synthesis.

6.7.5 Hearing tests

Other than the routine pure-tone audiometry where the peripheral auditory system is tested, there are a number of tests to evaluate the central auditory disorders and speech comprehension. Software for some of these tests is available (vagmionline.com). In *dichotic digit test*, pairs of digits are played to both the ears simultaneously. Figure 6.7 shows the waveform of a typical stereo dichotic signal. The client has to identify all the four digits in any order. The number of digits correctly recognized out of four gives a score for that stimulus. An average score of more than 80% has to be achieved for a normal hearing person. Dichotic digit test can be prepared in any language. The dichotic test is so

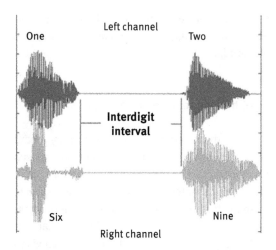

Figure 6.7: Waveform of a typical stereo signal used in a dichotic test. Competing digits are played simultaneously to both the ears (binaural). The client has to identify all the four digits.

powerful that it can identify simple cases such as dyslexia as well as complex cases such as tumors in the cortical regions.

Duration pattern test is meant to assess the ability to discriminate the difference in duration of tones (short vs long) having the same pitch. Based on the duration of the tones, six combinations are possible: short–long–short, short–long–long, short–short–short, long–long–short, long–long–long and long–short–short. One of the above six choices is picked up randomly and played over the headphones. The above six choices are displayed. The listener has to select one of the above as the response. In test mode, a score of at least 80% must be achieved. If not, it is a matter of concern.

Pitch patter test is meant to assess the ability to discriminate between a low-frequency tone and a high-frequency tone of the same duration. Based on the pitch, six combinations are possible as follows: low–high–low, low–high–high, low–low–low, high–high–low, high–high–high and high–low–low. One of the above six choices is picked up randomly and played over the headphones. The above six choices are displayed. The listener has to select one of the above as the response. In test mode, a score of at least 80% must be achieved. If not, it is a matter of concern.

Speech discrimination test is meant to assess the client's ability to discriminate between similar consonants such as "s" versus "sh" or "k" versus "g," "m" versus "n". In this test, a pair of words is played over the headphones. The listener has to identify if the consonant in the two words is the "Same" or

"Different." For example, the pair of words may be one of "she-see" or "she-she" or "see-see" or "see-she." In test mode, a score of at least 80% and above is considered normal. If the discrimination is poor, it can be improved using this module in "Training Mode." In training mode, if the answer is correct, a positive reinforcement is given. else, a message "listen carefully" is shown and that particular stimulus is replayed.

Speech intelligibility test evaluates a person's ability to identify a speech sound within a word, both for clean and for noisy or reverberant speech. Noise can be one of the various types such as white, babble, factory and traffic. Signal-to-noise ratio and the reverberation time can be chosen. In this test, a carrier phrase "Identify the word … " is used. A word with the target sound in the carrier phrase is played over the headphones. Several options of nearly similar sounding words are shown. The listener has to identify the word heard from the list of words displayed. For example, target sound: "s"; stimulus played: "sore." Words displayed for identification could be "pour," "core," "sore," "shore," "tore" or "four." In test mode, a score of at least 80% or above is considered normal.

6.7.6 Cognitive training module

Cognitive tests are meant for geriatric population to prevent degeneration of cognitive faculties. In order to prevent degeneration of brain tissues due to aging, researchers have found that the geriatric population must undertake stimulating tasks involving multimodal information processing. Tests are made in three modalities. *Auditory:* here digits in English are played binaurally; visual: abstract and meaningful pictures are used; *spatial:* the relative locations of an emerging pattern have to be remembered. A set of *23 cognitive tests or tasks* is available with gradually increasing order of complexity, referred to as *levels.* Tasks include forward/backward span, ascending/descending span, running span, math span, operation span, symmetry span, reading span, N-back, paired associate learning and pattern recognition memory.

6.8 Conclusion

In this chapter, we have covered the main communication disabilities. We have reviewed the speech production model and its associated parameters, digital speech signal processing techniques for speech analysis, synthesis and their relevance in the area of voice and speech disorders. Computer-assisted speech therapy today has proven to be useful and effective. We have given some examples of software solutions developed by the author that serve the needs of speech language pathologists, special teachers, therapists, voice specialists, ENT doctors and so on. Availing end user of such tools at remote locations and at a low cost requires a team effort by the funding agencies, NGOs, speech research scientists and speech technology engineers.

Future possibility in the area of signal processing seems to be very exciting. Software solutions, as described in this chapter, implement conventional signal processing algorithms, which attempt to find an explicit mathematical relationship between a speech signal and the desired parameter to be estimated. However, such a relationship is not available for all the parameters. For example, we do not have a mapping between a speech signal and the articulatory positions for consonants. There is a "paradigm shift" in the area of signal processing by a technology called "deep neural network" (DNN). Simply stated, a DNN attempts to mimic processing in a human nervous system. A live neuron receives inputs from several nervous paths, each input is given a certain weightage and all inputs are summed-up to obtain a value; the computed value is compared against a threshold and then the neuron either fires (generates a unit output) and remains neutral (generates no output). In a human brain there are billions of such neurons, which are interconnected in various combinatorial ways. In an artificial neural network, the basic functioning of a live neuron is mathematically modeled. In DNN, a very large number of neurons present at each of the several sequential layers are interconnected in all possible combinations. The weights of all such paths are determined using "training data." That is, samples of a speech signal are given as the input, and the estimated value of a parameter (by conventional methods) or directly and simultaneously measured value is specified at the output. The weights are adjusted till the DNN generates, as closely as possible, the desired known value of the parameter to be estimated. A large number of training samples (thousands) are used till the weights of all the paths converge and stabilize. Once a DNN is trained, then a new signal can be given as input, and the output of the trained DNN now corresponds to an estimate of the desired parameter. This approach does not require an explicit mathematical relationship between a speech signal and the desired parameter. Similar approach can be adopted for any desired parameter or any desired task. The use of DNN will revolutionize in finding a mapping between a speech signal and any desired parameter (say the position of an articulator) or any desired task (say, the type of voice and speech disorder). However, the only limitation at present is the requirement of a very large training data set and huge computational time.

Bibliography

[1]   Dictionary of communication disorders, Morris, Sec. Edn., Whurr Publishers, London (AITBS, India), 1994.

[2]   The voice and its disorders, Greene and Mathieson, Whurr Publishers, London (AITBS, India), 1995.

[3]   "Developmental phonological disorders and normal speech development", Pamela Grunwell, Vol 5, Issue 3, Oct 1989, Child Language Teaching and Therapy, Sage Publications.

[4]   Clinical measurements of speech and voice, R J Baken, Sec. Edn., Singular, 2000.

[5]   Is your voice telling on you? D. R. Boone, Plural Publishing, 2015.

[6]   Workshop on acoustic analysis – Summary statement, I R Titze, NCVS, 1995.

[7]   Invariance and variability in speech processes, J. S. Perkell and D. H. Klatt, Lawrence Earlbaum Associates, 1986. (Collection of papers.)

[8]   "Individual differences in vowel production", Keith J et al, JASA, 1994(2).

[9]   "Articulatory strategies, speech acoustics and variability" Espy Wilson, in From Sound to Sense 50+ years of discoveries in speech communication, MIT Press, 2004.

[10]   "DSP for voice and speech disorders and enhancement", T V Ananthapadmanabha, in 'Signal Processing for Speech and Hearing Disorders', WiSSAP 19–22 January 2018, IITG, Guwahati.

[11]   "Significance of spectral valleys with application to front-back distinction of vowel sounds", T V Ananthapadmanabha, A G Ramakrishnan and Shubham Sharma, Arxiv 1506.04828.

[12]   "Intrinsic-cum-extrinsic normalization of formant data of vowels", T V Ananthapadmanabha and A G Ramakrishnan, JASA-EL2016. Also see Arxiv 1609.05104 for a detailed presentation.

[13]   "Estimation of the vocal tract length based on the frequency of the significant spectral valley", T V Ananthapadmanabha and A G Ramakrishnan, Inter Speech Conference, 2018.

Chitralekha Bhat, Anjali Kant, Bhavik Vachhani, Sarita Rautara and Sunil Kopparapu

7 A mobile phone-based platform for asynchronous speech therapy

Abstract: Traditionally, rehabilitation of persons with speech disabilities has been conducted solely face to face, either physical or virtual interactions between the therapy provider and the person in need of therapy. Interdisciplinary and collaborative work between speech language pathologists (SLP) and speech technologists has paved way to tools and techniques that enable rehabilitation without having to be face to face (asynchronous). Such tools can significantly impact the therapy efficiency especially in India, considering the infrastructure limitations, language variability, lack of experts and awareness on one side as well as a deep penetration of mobile phones on the other side. In this chapter, we describe the process adopted in building an asynchronous mobile phone-enabled speech therapy platform that reduces the number of face-to-face interactions with SLP, yet increases the efficiency of the therapy process by engaging the patient more frequently, asynchronously and remotely. The essential components of the platform are (a) a web interface for the SLP to assign therapy exercises and monitor the progress of their patients at any time round the clock and (b) a mobile application for the patient to practice the personalized therapy exercises at a time and location of their convenience. Speech processing algorithms running on the mobile device perform objective and automatic assessment of the patient speech and provide actionable feedback to the patient in real time.

Keywords: Remote speech therapy, misarticulation, automatic assessment, speech disorders, asynchronous speech therapy

7.1 Introduction

Internet of things (IoT) and pervasive frameworks involving smart devices have revolutionized the healthcare domain [1, 2]. Rehabilitation after a medical condition or ambient assisted living for the elderly see a shortage of healthcare

Chitralekha Bhat, Sunil Kopparapu, TCS Research and Innovation, Mumbai, India
Anjali Kant, Sarita Rautara, Ali Yavar Jung National Institute for Speech and Hearing Disabilities, Mumbai, India
Bhavik Vachhani, ScribeTech, Mumbai, India

https://doi.org/10.1515/9781501501265-008

professionals and such scenarios are typical applications for smart IoT-based systems [3]. Speech analysis research is gearing up toward building smart and automated systems for rehabilitation of patients with speech and hearing disorders. Serious attempts are being made toward understanding the application of speech signal processing and machine learning techniques to pathological speech [4]. Automatic assessment of pathological speech has been studied extensively by speech researchers [5, 6]. Automatic speech recognition (ASR) technology in conjunction with multimedia gaming has been used to provide computer-based therapy to patients with dysarthric speech [7]. CloudCAST project [8] is an example of a cloud-based speech rehabilitation system created with the objective of providing help to professionals who work with individuals having speech disorders, with a keen user-centric design. Malavasi et al. [9] describe a low-cost implementation of the CloudCAST platform to efficiently interact both with the traditional home automation systems and the IoT solutions, using mobile devices.

Mobile device-based rehabilitation is seen as a winner for healthcare especially with conditions such as limited infrastructure, limited available experts in speech pathology, lack of awareness, language variability on one side and deep penetration and low cost of the mobile phones on the other side. This scenario demands a low-cost and sustainable solution that will address the skewed ratio of speech language pathologists (SLPs) to patients in a language-independent manner. As per the census of 2011, 7.5% of the total disabled population comprises persons with speech disabilities as shown in Figure 7.1. Additionally, general-purpose automatic speech recognizers or smart speakers do not often cater to persons with speech disabilities, mandating development of specialized speech technology for the purpose of assistive technology.

Figure 7.1: Types of disability (in %) in India as per census 2011 [10].

In this chapter, we describe our approach toward creating an asynchronous mobile-based assistive platform for speech rehabilitation. Platform showcases an interdisciplinary approach to rehabilitate patients with speech sound disorders. The novelty of this platform lies in the fact that the conventional therapy process is augmented using automatic speech processing algorithms to enable instant feedback to the patient, while retaining the SLP involvement and intervention at regular intervals which is crucial in a therapy process. Automatic assessment of pathological speech using speech signal processing is an integral part of this smart system. The essential components of the platform are (i) a web interface for the SLP to assign therapy exercises and monitor the progress of their patients at any time round the clock and (ii) a mobile application for the patient to practice the personalized therapy exercises at a time and location of their convenience. Major contributions of this chapter are (a) digitization of the existing paper-based assessments and therapy drills to enable availability of electronic medical data, (b) personalization to suit specific kind of target audience, (c) development of speech processing algorithms for objective assessment of patient speech and (d) evaluation of the platform with five patients with articulation errors in their speech.

7.2 Digitization of assessments and therapy drills

Sounds in conversational speech are produced by means of a sequence of articulator movements, requiring exact placement, sequencing, timing, direction and force of the articulators. This sequence occurs simultaneously with precise air-stream alteration, initiation or halting of phonation and velopharyngeal action in order to produce speech sounds. Assessment of speech sound production therefore requires considerable skill and knowledge. During the initial stages of the language acquisition process, children make mistakes such as omission or substitution of consonants; however, a child is expected to produce sound correctly by a stipulated age as shown in Figure 7.2. If a person fails to acquire the speech sounds within an acceptable time frame or is unable to pronounce the sound correctly within a sentence or a word for every repetition, then this person is considered to have an articulation disorder in speech. Articulation disorder may be defined as the difficulty in producing a single or a few sounds with no pattern or derivable rule. The cause of the articulation error could be cleft lip and palate or developmental or due to hearing impairment and others. Prolonged and severe articulation problems call for SLP's attention. Focus of the therapy is on identifying the articulation errors and developing a personalized drill to enable correction of misarticulated sounds.

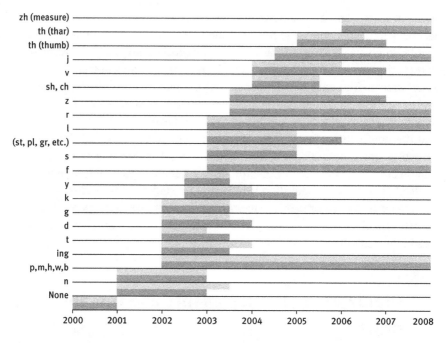

Figure 7.2: Language acquisition in children [11] assumed born in the year 2000. (boy dark gray, girl light gray).

7.2.1 Traditional approach

Articulation disorder is ascertained by administering the photo articulation test (PAT) preceded by a physical examination of the patient, together known as the pretherapy procedure. PAT [12] in its current form is a collection of photographs in black and white, accompanied by case sheets to record the patient details and the manual analysis of patient speech for each word. Each picture within the PAT is required to elicit a word corresponding to a speech sound at a specific position such as word initial, middle or ending. For example, Figure 7.3(b) is supposed to elicit the word saaykal (cycle) corresponding to the sound /k/ while Figure 7.3(c) is supposed to elicit the word maachi (fish) corresponding to the sound /m/. When administered for all phonemes in Marathi, the total number of unique picture–word pairs amount to 108. A patient's speech is tagged as either *correct, substitution, omission, addition or distortion* by the SLP based on perceptual observations. Certain drawbacks of the traditional method are (1) it is cumbersome for the SLP to administer the test as it does not allow for any customization for a specific patient, (2) cannot be

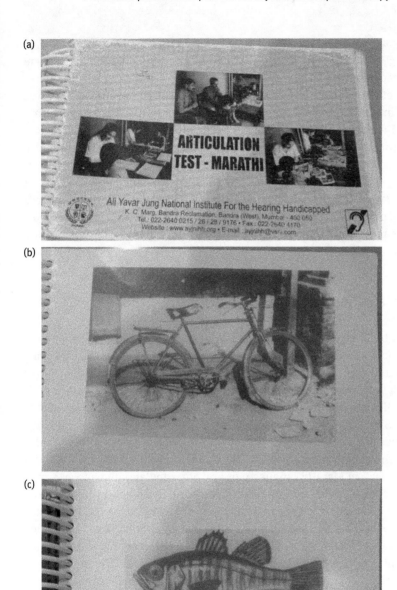

Figure 7.3: (a) Marathi PAT manual and (b, c) pictures from the manual [12].

selectively administered for sounds in error during subsequent visits, (3) drab and boring for the patient and (4) difficult to track the analysis and patient (manual) records.

The pretherapy procedure was digitized through a user-centric approach as described in [13]. It was built as a mobile application so that it is portable, easy to use and engaging, especially for the child patients. At this point, a digital, contemporary, color picture was used to represent each picture–word pair in the PAT. The patient's response to a picture as well as SLP's subjective assessment could be recorded on the mobile device. An evaluation of PAT (D-PAT) for its ease of use while still achieving the goals of PAT showed majority of patients as well as the SLP's preferred D-PAT to the traditional book-based method. Moparest's pretherapy user guide is available at https://sites.google.com/site/projectmoparest/publications.

7.2.2 Data collection

In order to build algorithms that can automatically assess speech it is imperative to understand the underlying speech production mechanism, especially for disordered speech. The perceptual, subjective assessments of the trained SLP's need to be replicated as closely as possible. Note that a human expert is trained to listen to a variety of speech and acquires the skill through ample examples. Similarly, a machine needs to be trained with examples of both disordered speech and normal speech to be able to provide an assessment that is closest to the subjective assessment. No standard speech database exists for children speech in Marathi language, or any other Indian language as a matter of fact. Hence, collection of speech data is a crucial step in the development of the assessment tool. Speech data corresponding to the 108 PAT words was collected using a home-grown data collection application, which was build along the lines of D-PAT. A total of 50 school children with normal speech participated in this exercise. Age of children ranged between 6 and 12 years. Since the objective was to capture the correct pronunciations in Marathi language, care was taken that each child's mother tongue was Marathi. The activity was conducted over a period of a week wherein data was collected on the school premises. The environment was found to be significantly noisy. The data thus acquired was then collated and hand labeled to be used for training the speech algorithms.

7.2.3 Therapy exercises

The objective of therapy exercises is to provide a patient enough opportunity to practice the speech sound in which the patient has articulation error. It was deemed that a minimum of 10 therapy words is essential for each phoneme (speech sound) at each of the word positions, namely, *initial, middle* and *final.* Hence for each phoneme, a minimum of 30 picture–word pairs are necessary in order for the therapy exercises to make an impact. Based on the prevalence, six Marathi consonants, namely, /k/, /g/, /t/, /d/, /s/ and /sh/ were chosen during the first phase of this work. For designing the therapy exercises, common and familiar words that could be represented through pictures were selected from a set of 500 words that were sourced from different sources, and five SLPs were asked to comment on the usability of the picturable words in terms of (a) appropriateness of the selected words and the corresponding pictures with respect to the target/intended phoneme which they represent, and (b) their judgment of the word being familiar to a 4-year-old child. Their response was evaluated to finalize the set of pictures associated with the phoneme and its position of articulation. This selected list of picture–word pairs was further validated by 100 children. The entire process is described in detail in [14]. This resulted in the final list of therapy exercise word–picture pairs suitable to represent the six consonants in Marathi in all the three positions. These pictures were then incorporated into the mobile application described in Section 7.3 to be used as practice exercises. In case of articulation errors in multiple phonemes, therapy practice is done sequentially (as conducted in the *face-to-face* therapy sessions), wherein the sequence of the phonemes is decided by the SLP. One of the objectives of the platform is to also provide instant feedback to the patients regarding their articulation. This is made possible by using speech processing algorithms specifically designed to automatically analyze speech with articulation errors, recorded using a mobile phone microphone. Speech algorithms built for this purpose are described in the next section.

7.3 Platform design consideration

We describe the broad framework that was used to build the platform. This process is independent of the language of therapy and the nature of speech disorder being treated. The overall design process remains the same when the language changes. The drill lessons and the instructions to the patients will be custom adapted to the specified language. While the design of the therapy process changes including the *face-to-face* therapy session pattern, drill lesson detailing and the mobile application may differ from disorder to disorder, and the overall design process remains the same. Figure 7.4 shows the key steps involved in designing the platform. It is crucial that the *face-to-face* therapy be adapted to digitization in such a way as to both create an impact for the patient and SLP as well as not lose the integrity and benefits of the *face-to-face* therapy.

Figure 7.4: Platform design process.

Figure 7.5 shows the system architecture for the platform. Each component of the platform is described in subsequent sections. The objective of this platform is to reduce the therapy time and cost by efficiently using the SLP's time and expertise. Speech therapy requires constant monitoring and interventions by the SLP in order for the therapy to succeed. However, rural Indian patients who are in need of speech therapy fail to maintain regular attendance of *face-to-face* sessions with the SLP, mainly due to the travel and opportunity costs involved and a lack of awareness. Also, they are either unable to or do not practice the exercises or drills that are prescribed by the SLP. The infrequent visits and lack of supervised practice sessions impact the therapy process adversely, leading to the inevitable regression of patients, consequently prolonging the duration of the therapy process. The platform enables continuous and supervised training for the patients. A typical life cycle of therapy using the platform is shown in Figure 7.6.

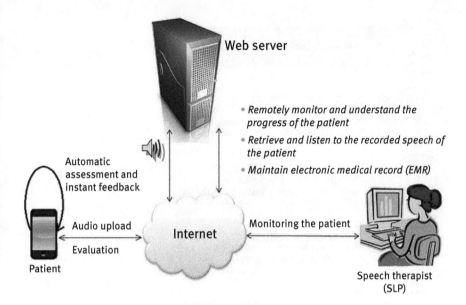

Figure 7.5: System architecture of the platform.

7.3.1 SLP interface

The web interface is the SLP's touch point to the platform. The primary purpose of the SLP interface is to allow SLPs to (a) enroll a patient, (b) assign drills and practice sessions to the patient and (c) monitor the progress of the patient. The SLP can enroll a patient into the platform through the web interface. Personal information such as age, gender, mother tongue is captured as part of the enrollment process. This information is used for personalization of the mobile application as well as for improving the speech analysis process. During the enrollment, the SLP selects one or more phoneme or speech sound that needs to be practiced. In case of the patient requiring multiple phoneme correction, the order in which they need to be practiced is also selected by the SLP. In this way the web interface allows the SLP to design the therapy session; since they can continuously monitor the progress, the web interface allows for frequent updating of the therapy sessions by the SLP. The cause of articulation disorder, the severity of the disorder and the estimated time, determined by the SLP, required to complete therapy process are also captured by the platform. Personal and therapy information of a patient is saved into the platform database as part of the electronic medical record (EMR) database. A sample snapshot of patient monitor page is shown in Figure 7.7. Details of the patient are on the left panel while the exercise

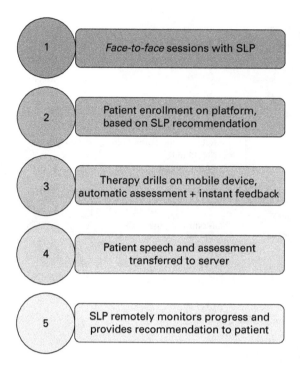

Figure 7.6: Typical therapy life cycle.

practiced by the patient is on the right. The snapshot also shows the practiced drills corresponding to the phoneme /k/ in the word *middle* and word *final* positions by the patient. The web interface enables the SLP to play the audio and listen to the patient's utterance of a specific word; this also allows them to check, and if required correct, the automatic evaluation (see Figure 7.7). Note that we have selected three kinds of errors, namely, substitution, omission and distortion. Based on the audio, the SLP can make his/her own subjective assessment and mark the practice exercise as such through a drop-down menu.

7.3.2 Mobile application

The mobile application is designed to enable a patient to practice therapy drills as planned by the SLP. Post enrollment, the patient is provided login credentials for the mobile application. Patient's information on the platform database is synchronized with the mobile application using the patient login credentials.

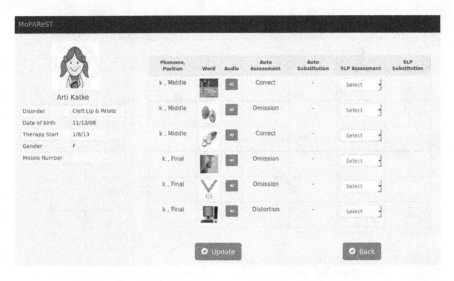

Figure 7.7: The platform–web interface.

In order for the patient to practice drill lessons for a particular phoneme, visual stimuli are provided for words with the error phoneme in the *initial, middle* or *final* positions of the word. Note that the web console gives the SLP scope to view the performance of the patient and also if required make any changes to the therapy drill. Any changes in the therapy drill sessions are automatically pushed to the mobile application of patients. Figure 7.8 shows a view of the mobile application for the phoneme /s/ in *initial, middle* and *final* positions.

Patient needs to identify the word from the visual stimulus, speak the word while recording it and then submit the utterance for automatic evaluation. The valuation happens on the mobile device itself using speech signal processing techniques described later in Section 7.4. It is important to note that no Internet access is needed to run the speech algorithm. An instant visual feedback indicating whether the patient spoke the phoneme correctly or not is provided. Multiple attempts to utter the same word are allowed based on the assessment by the speech algorithm, and the number of attempts is configurable. An audio cue (with the correct articulation of the word) for every word and an instructional video in Marathi language (describing how a phoneme should be articulated) for every phoneme are also provided to help the patient with the therapy drills. Instructional videos are in local language and provide exactly the same instructions provided by the SLP during the *face-to-face* establishment session. The drill itself is divided into multiple levels and the patient needs to complete the drill exercises level-wise. A patient can

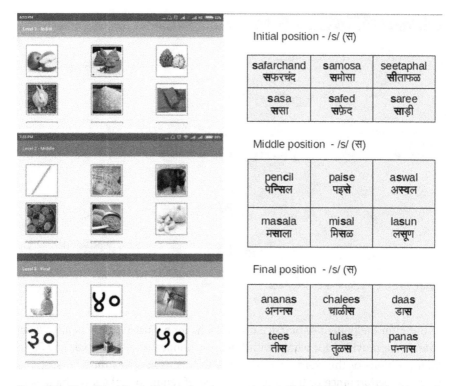

Initial position - /s/ (स)

safarchand सफरचंद	samosa समोसा	seetaphal सीताफळ
sasa ससा	safed सफ़ेद	saree साड़ी

Middle position - /s/ (स)

pencil पेन्सिल	paise पइसे	aswal अस्वल
masala मसाला	misal मिसळ	lasun लसूण

Final position - /s/ (स)

ananas अननस	chalees चाळीस	daas डास
tees तीस	tulas तूळस	panas पन्नास

Figure 7.8: View of mobile application for phoneme */s/* at *initial*, *middle* and *final* positions of the stimulus.

progress to the next level *Level*$_{next}$ based on the scores (*score*$_i$) of the current level (*Level*$_i$). This check is required to ensure that the patient is practicing as desired. The progression of a patient currently in *Level*$_i$ based on his score *score*$_i$ (determined by the speech processing algorithm) is shown in Algorithm 7.1, where T_q is the qualifying score to proceed to the next level and T_m is the minimum score below which the patient needs to visit the SLP. The thresholds T_q and T_m are configurable.

The multiple levels and instant feedback simulate a game-like environment that appeals to the young patients. Figure 7.9 shows the screen for the Marathi word *kulup* (lock) corresponding to phoneme */k/* in the *initial* position.

In its current form the platform caters to the correction of articulation errors in Marathi language, wherein a patient is unable to pronounce a particular sound correctly. The SLP first establishes the speech sound that needs correction during the *face-to-face* sessions, ensuring that the patient has understood how to

Data: *Level$_i$, score$_i$*
Result: *Level$_{next}$*
initialization;
while *Level$_i$* **do**
    read *score$_i$*;
    if *score$_i$* ≥ *T$_q$* **then**
        Level$_{next}$ = *Level$_{i+1}$*;
        Go to next level;
    end
    if *T$_q$* > *score$_i$* ≥ *T$_m$* **then**
        Level$_{next}$ = *Level$_i$*;
        Play instructional video;
    end
    if *score$_i$* < *T$_m$* **then**
        Level$_{next}$ = *Level$_i$*;
        Ask to meet the SLP;
    end
end

Algorithm 7.1: Monitoring progress of a patient.

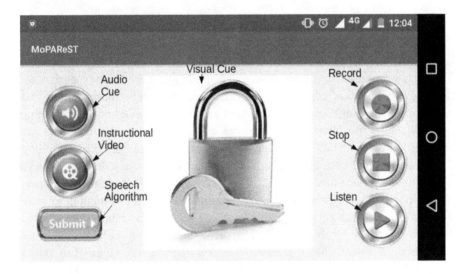

Figure 7.9: Platform – patient application screenshot.

pronounce the speech sound correctly. The patient is then transferred onto the platform for further practice via drills. For each speech sound, we define a specific repertoire of drills to be followed on the platform based on SLP's inputs. These drills are built into the mobile application, which is used by the patient.

7.4 Speech algorithms

Several aspects need to be considered when building an automatic speech analysis algorithm

– Channel and specifications of the recording environment such as sampling rate, audio format and background noise [15]
– Target audience – to build appropriate acoustic models (AM) [16]
– Target speech language – to build appropriate language models (LM) [16]

All the above aspects were dealt with to achieve high correspondence between automatic speech assessment from the speech algorithm and the perceptual subjective assessment of an SLP. Also, since we envision this platform to be used extensively by rural Indian patients, the speech algorithm should work without Internet access. Sphinx-4 [17] ASR engine was configured within the mobile application such that the ASR runs on the mobile device itself with no Internet requirement.

7.4.1 Noise robust processing of mobile phone speech

In [15], we propose a multitaper-based perceptual linear prediction speech processing front end to achieve robust phonetic segmentation for mobile phone and tablet recordings of speech contaminated with environmental noise or clipping or both. Multitaper-based speech features are robust to both channel noise and background noise. Traditionally, speech features are extracted by transforming the speech signal from the temporal domain to the frequency domain through short-time Fourier transform using Hamming window and a single taper spectral estimation. In this chapter, we use multitaper spectral estimation to compute Mel frequency cepstral coefficients (MFCC) for the mobile phone-based speech recordings of the patients. The details of applying multitaper spectral estimation to speech features have been described in [15]. Figure 7.10 shows how a typical MFCC feature extraction is modified using a multitaper spectral estimation. Twelve-dimensional MFCC features were computed using Thomson multitaper spectral estimation with a 30 ms window and a 10 ms shift rate.

7.4.2 Articulation error identification

We describe how automatic analysis of speech was used to assess the patient speech for articulation errors. Speech features or parameters, AMs and LMs

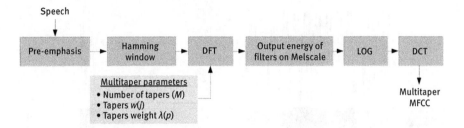

Figure 7.10: Multitaper MFCC speech front end.

play a key role in the performance of an ASR. Our focus is on assessment of patient speech to yield, for a given phoneme, one of the following categories: *correct* (articulated the phoneme correctly), *substitution* (replaces with another phoneme), *omission* (does not articulate the phoneme) or *distortion* as done in the traditional approach described in Section 7.2. AM and LM are engineered to achieve this goal.

7.4.2.1 Acoustic models

Target audience for this application is typically of the age 4–12 years. However, the application needs to cater to patients of all age ranges. A Hindi (similar phone-set as of Marathi) speech corpus [18], of adult speakers with normal speech, was used to build the initial AM. These AMs were then adapted using the normal speech of 50 children collected as described in Section 7.2.2. An element of personalization based on the patient's age was used to choose the AM to be used for articulation assessment.

The phonetic characteristics of a consonant were used to fine-tune the AM [16]. As shown in Figure 7.11, a consonant or a stop comprises two regions, namely *closure* (or voice bar) for an unvoiced (or voiced) consonant, followed by a *burst* [19]. Also studies show that the spectral characteristics of the consonants are not dependent on the preceding or following vowel [20], which motivates us to inspect the stop consonants as a stand-alone sound. Traditionally, AMs consider the *closure* and *burst* regions separately (initial AM) during training. We propose training of consonants using region of *closure* and the following *burst* as one unit (cluster AM) as shown in Figure 7.11. Stop cluster AMs are used for automatic assessment of articulation [16].

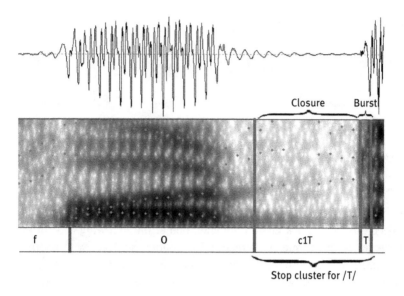

Figure 7.11: Stop cluster for the consonant /t/.

7.4.2.2 Language model

Figure 7.12 shows the phone chart for consonants in Hindi (similar as Marathi). In this chart, the consonants are arranged horizontally based on voicing and aspiration, and vertically, based on place of articulation (PoA) and manner of articulation (MoA).

We design an LM that caters to *correct*, *substitution* and *omission* types of articulations. When we analyze an utterance for *substitution* type of articulation error, it is important to understand the typical errors that patients make. The possible substitutions are either due to a change in PoA, MoA or due to change in the voicing parameter. The possible substitutions for /k/, /g/, /t/ and /d/ are marked in Figure 7.12. An example for the Hindi word *gaay* is shown in Figure 7.12. The task of the ASR is to identify the consonant being spoken, wherein the consonant that needs to be spoken is known a priori. We consider that the utterance is erroneous if the consonant being spoken (as recognized by ASR) is different from the one expected. In order to give actionable feedback to the patient, such as corrections to the articulator positions, ASR needs to be able to accurately identify the consonant being spoken. For each utterance, we build a set of pronunciations to form the finite state grammar (FSG), based on prior knowledge of the possible articulation errors. This FSG is the LM and consists of the canonical form for each

Voicing Place manner	Unvoiced unaspirated	Unvoiced aspirated	Voiced unaspirated	Voiced aspirated	Nasal
Velar stop-plosive	[k] क	[kh] ख	[g] ग	[gh] घ	
Coronal-alveolar affricate	[c] च	[ch] छ	[j] ज	[jh] झ	
Coronal-alveolar (or rretroflex) Stop-plosive	[T] ट	[Th] ठ	[D] ड	[Dh] ढ	[N] ण
Interdental fricative	[t] त	[th] थ	[d] द	[dh] ध	[n] न
Bilabial stop-plosive	[p]	[ph] फ	[b] ब	[bh] भ	[m] म
Glide			[y], [w] य , व		
Liquids (Lateral + Rhotic)			[r], [l] र , ल		
Coronal postalveolar fricative	[S], [S] श , स	[h] ह			

Figure 7.12: Hindi Phone chart for consonants.

utterance and alternative pronunciations as per the substitutions in Figure 7.12. An example of the LM creation for the word *kaan* in Hindi, where /k/ is the target consonant at initial position, is shown in Figures 7.13 and 7.14.

```
#JSGF V0.1;
grammar RST_database;
public <basicCmd> = <object>;

<object> = (k A nn |g A bb |tt A nn );
```

Figure 7.13: Variants for the word *kaan*.

Thus, the speech algorithm used to automatically assess the patient speech for articulation errors is specialized to cater to both adult and children speech

Phone (/g/) position	Actual word	MOA/POA error	Voicing error
Initial	गाय /g aa y/	काय /k aa y/	दाय /d aa y/
Middle	बैंगन /b ey n g a n/	बैंकन /b ey n k a n/	बैंदन /b ey n d a n/
Final	पतंग /p a t a n g/	पतंक् /p a t a n k/	पतंद् /p a t a n d/

Figure 7.14: Words in Hindi PAT for consonant /g/.

recorded using mobile phone microphones. This is an integral part of the platform, wherein the instant feedback and progression of the game-like practice on mobile phones is based on the outcome of the speech algorithm.

7.5 Conclusions

Pervasive and smart systems are playing a key role in the healthcare domain, specifically in rehabilitation. This chapter describes one such framework for providing speech therapy remotely and asynchronously. Aspects regarding the key components of the platform such as the web interface for SLPs, mobile application for patients and the underlying speech algorithm for automatic evaluation of the patient are elaborated. Evaluation of this platform is being done for five patients with articulation errors in Marathi language. The platform framework can be extended to provide speech therapy to multiple speech sound disorders in many different languages. Further, the system also serves as EMR for patient progress.

Acknowledgments: This research, called MoPAReST (Mobile Phone-Assisted Remote Speech Therapy), was supported in part by the Science for Equity Empowerment and Development (SEED), Department of Science and Technology, Government of India (SEED/TIDE/013/2012).

References

[1]   Islam, S.M.R., Kwak, D., Kabir, M.H., Hossain, M., and Kwak, K.S. The internet of things for health care: A comprehensive survey. IEEE Access, 3, 678–708, 2015.

[2]   Pang, Z. Technologies and Architectures of the Internet-of-Things (IoT) for Health and Well-being, Ph.D. thesis, KTH Royal Institute of Technology, 2013.

[3]   Fan, Y.J., Yin, Y.H., Xu, L.D., Zeng, Y., and Wu, F. IoT-based smart rehabilitation system. IEEE Transactions on Industrial Informatics, 10(2), 1568–1577, May 2014.

[4]   Gupta, R., Chaspari, T., Kim, J., Kumar, N., Bone, D., and Narayanan, S.S. Pathological speech processing: State-of-the-art, current challenges, and future directions. In Proc (ICASSP), Mar 2016.

[5]   Middag, C., Martens, J.P., Van Nuffelen, G., and De Bodt, M. Automated intelligibility assessment of pathological speech using phonological features. EURASIP J. Adv. Signal Process, 3(1–3), 9, Jan 2009.

[6]   Maier, A., Haderlein, T., Eysholdt, U., Rosanowski, F., Batliner, A., Schuster, M., and Nöth, E. PEAKS A system for the automatic evaluation of voice and speech disorders. Speech Communication, 51(5), 425–437, 2009.

[7]   Ganzeboom, M., Yilmaz, E., Cucchiarini, C., and Strik, H. On the development of an ASR-based multimedia game for speech therapy: Preliminary results. In Proc ACM Workshop, 3–8, MMHealth '16, 2016.

[8]   Green, P.D., Marxer, R., Cunningham, S.P., Christensen, H., Rudzicz, F., Yancheva, M., Coy, A., Malavasi, M., Desideri, L., and Tamburini, F. Cloudcast – remote speech technology for speech professionals. In: INTERSPEECH (2016)

[9]   Malavasi, M., Turri, E., Atria, J., Christensen, H., Marxer, R., Desideri, L., Coy, A., Tamburini, F., and Green, P. An innovative speech-based user interface for smart-homes and IoT solutions to help people with speech and motor disabilities. Studies in Health Technology and Informatics, 242, 306–313, September 2017.

[10]   CensusCommissioner: Census 2011, Disabled Population India. http://censusindia.gov. in/Census_And_You/disabled_population.aspx, viewed May 2019

[11]   Smit, A.B., Hand, L., Freilinger, J.J., Bernthal, J.E., and Bird, A. The Iowa articulation norms project and its Nebraska replication. Journal of Speech and Hearing Disorders, 55 (4), 779–798, 1990.

[12]   AYJNIHH, RRTC: Articulation Test – Marathi. Ali Yavar Jung National Institute of the Hearing Impaired and Regional Rehabilitation Training Center (1993)

[13]   Bhat, C., Chaplot, A., Kant, A., and Kopparapu, S.K. Digitization of Hindi photo articulation test for speech sound disorders. In: 4th International Workshop on User-Centered Design of Pervasive Healthcare Applications. ICST (July 2014)

[14]   Sinha, A., and Kant, A. Development of a digital based test of articulation in Marathi. Journal of Communication Disorders of AYJNISHD(D), 1, 2016.

[15]   Vachhani, B., Bhat, C., and Kopparapu, S. Robust phonetic segmentation using multi-taper spectral estimation for noisy and clipped speech. In: 2016 24th European Signal Processing Conference (EUSIPCO). pp. 1343–1347 (Aug 2016.

[16]   Bhat, C., Vachhani, B., and Kopparapu, S. Automatic assessment of articulation errors in Hindi speech at phone level. In: TENCON 2015-2015 IEEE Region 10 Conference. pp. 1–4 (Nov 2015)

[17]   Walker, W., Lamere, P., Kwok, P., Raj, B., Singh, R., Gouvea, E., Wolf, P., and Woelfel, J. Sphinx-4: A flexible open source framework for speech recognition. Tech. rep., Sun Microsystems, Inc, Mountain View, CA, USA, 2004.

[18]   AwazYP: https://sites.google.com/site/awazyp/data/speechcorpus, viewed April 2019.

[19]   Stevens, K.N. Models for the production and acoustics of stop consonants. Speech Communication, 13(34), 367–375, 1993. http://www.sciencedirect.com/science/article/pii/016763939390035J.

[20]   Blumstein, S.E., and Stevens, K.N. Acoustic invariance in speech production: Evidence from measurements of the spectral characteristics of stop consonants. The Journal of the Acoustical Society of America, 66(4), 1001–1017, 1979. http://scitation.aip.org/content/asa/journal/jasa/66/4/10.1121/1.383319.